To Demis,   April 2021

Thanks for all
of your good advice
and support.

Jay

Geometric Foundations of Design

Old and New

Geometric Foundations of Design

Old and New

Jay M. Kappraff

New Jersey Institute of Technology, USA

World Scientific

NEW JERSEY · LONDON · SINGAPORE · BEIJING · SHANGHAI · HONG KONG · TAIPEI · CHENNAI · TOKYO

Published by

World Scientific Publishing Co. Pte. Ltd.

5 Toh Tuck Link, Singapore 596224

USA office: 27 Warren Street, Suite 401-402, Hackensack, NJ 07601

UK office: 57 Shelton Street, Covent Garden, London WC2H 9HE

Library of Congress Cataloging-in-Publication Data

Names: Kappraff, Jay, author.
Title: Geometric foundations of design : old and new / Jay Kappraff,
 New Jersey Institute of Technology, USA.
Description: New Jersey : World Scientific, [2021] | Includes bibliographical references and index.
Identifiers: LCCN 2021003622 (print) | LCCN 2021003623 (ebook) |
 ISBN 9789811219702 (hardcover) | ISBN 9789811219719 (ebook) |
 ISBN 9789811219726 (ebook other)
Subjects: LCSH: Geometry. | Mathematics. | Design.
Classification: LCC QA447 .K375 2021 (print) | LCC QA447 (ebook) | DDC 745.401/516--dc23
LC record available at https://lccn.loc.gov/2021003622
LC ebook record available at https://lccn.loc.gov/2021003623

British Library Cataloguing-in-Publication Data
A catalogue record for this book is available from the British Library.

For any available supplementary material, please visit
https://www.worldscientific.com/worldscibooks/10.1142/11811#t=suppl

Typeset by Stallion Press
Email: enquiries@stallionpress.com

Desk Editor: Soh Jing Wen

Printed in Singapore

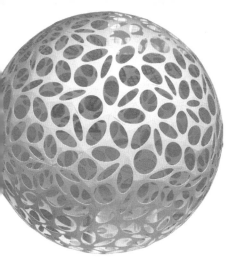

Contents

Chapter 10. Why is a Donut Like a Coffee Cup?: An Introduction to Topology   107

Chapter 11. The Szilassi and Csaszar Polyhedra   127

Chapter 12. Curves of Constant Width and a Three Dimensional Sculpture that Rolls   133

Chapter 13. An Introduction to Fractals   141

Chapter 14. Creating a Fractal Wallhanging   149

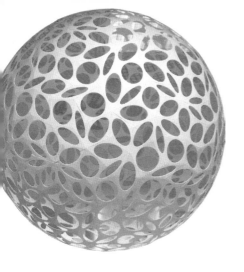

Cover Design Description

1. The Front Cover

The design on the front cover of this book, created by Werner Van Hoeydonck, a Belgian architect and ornamental designer based in Vienna, is entitled "The Ambassadors — Nr. 1". It was one of 27 designs inspired by geometric motifs in the Hall of the Ambassadors, the largest room in the Alhambra, a palace and fortress complex from 889 AD in Granada, Spain. I feel that this design is very much in the spirit of this book which is why I chose it for the front cover.

Listen to what Werner has to say about his designs:

"They "speak" a more universal formal language, and are, like music, understood instinctively by people around the world. It is as if the architect wanted to make a statement by placing two specific geometric motifs in the most important symbolic locations within the Hall of the Ambassadors and made them in a purely geometric, "Platonic," style totally different from the other motifs in the room. They were the starting point for this series of designs. They are "ambassadors", a visualization of a safe haven for people who work for good relations and connections between nations and people around the world.

They should be seen as symbols of "enspacement," of development in space symbolize growth, harmony, and unity. These works all have a strong focus; they lead viewers to "center" themselves. They are "ambassadors" for the universal message of love and peace."

The Ambassadors are available in three printing techniques: Silverprint, Chromaluxe, or Diasec.

For more information and special requests: www.wernervanhoeydonck.com

2. The Back Cover

The image on the back cover was constructed by Haresh Lalvani, an internationally recognized architect-morphologist, artist-inventor and design scientist at Pratt Institute. Chapter 23 describes how he built a Hypersurface series of sculptures from flat metal sheets [Lal7] Three examples from this series are a sphere, an elongated ellipsoid and an oblate ellipsoid. The back cover of the book features what Lalvani refers to as HOLEYSPHERE (2007), painted steel, laser-cut, 24" diameter. Holeysphere was displayed by Moss gallery in their New York gallery and at Design Miami 2011.

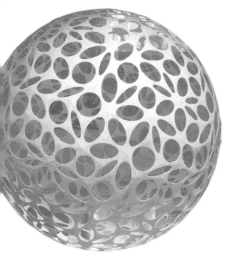

Preface

Geometric Foundations of Design: Old and New is my fourth book on the subject of mathematics and design. My first book, *Connections: The Geometric Bridge between Art and Science*, was written to set the limits on a new discipline that was dawning in 1990. This new area of study was to be thoroughly interdisciplinary, bringing mathematics, science, architecture, computer science, art and design together under one roof. The book was also dedicated to the many non-academics who were working on these ideas outside of the academy: crafts people, computer artists and amateur mathematicians. In my next book, *Beyond Measure: A Guided Tour through Nature, Myth and Number*, I attempted to show how many of the ideas of modern art and design had roots in ancient or very old themes; there was an historical and cultural basis for this new discipline. My third book, *A Participatory Approach to Modern Geometry*, is a textbook on geometry that seeks to pay homage to geometry which is at the root of all mathematical thought and, in particular, this new discipline. It sought to remove geometry from its constraints as a subject based solely on theorems and proofs, merely an exercise in pure logic and a subject of interest solely to mathematicians. It opened this discipline to the world to show that it is connected to the whole of human endeavor. This fourth book continues to expand the frontiers of this new discipline by inviting the reader to participate in the creation of their own designs while gaining an appreciation of how mathematics and design are deeply intertwined. Accompanying this text is a website (`https://www.worldscientific.com/worldscibooks/10.1142/11811#t=suppl`) in which many of my students' designs are shown in color along with miscellaneous information. The website is open source and free to download. There will also be an opportunity to expand the book beyond its present timeline.

I began to develop my ideas about the relationship of mathematics and design in 1970 through my friendship with Mary Blade who was teaching math and design to students from the School of Architecture at Cooper Union College for more than 25 years. Following her lead, I taught a course on the mathematics of design for many years at New Jersey Institute of Technology (NJIT) to students from the College of Architecture and Design. Through teaching this course, it became clear to me that large ideas from the realm of geometry come forth from humble beginnings. I have written this book to give voice to this insight. Many of the topics in the 26 chapters of this book were inspired by ancient mathematics and also the craft of artists and designers. In this introduction, I have attempted to show how designs motivate rich mathematical ideas. This project began with my friendship with, Slavik Jablan, a scholar of symmetry, knot theory, and the history of design from the University of Belgrade and the editor of the Journal of Visual Mathematics. Slavik was also a fine artist who created a large oeuvre of oil paintings and watercolors as well as computer art. Unfortunately, Slavik's untimely death in 2017 prevented our collaboration. I have included many of his ideas. You will also find materials written by his student, Ljiljana Radovic, who is now a Professor on the faculty of Mechanical Engineering at the University of Nis in Serbia.

The book can be read as a collection of topics in geometry. To a great degree, each chapter can stand alone, although there is some attempt to find a logical order. Each chapter has several constructions to be carried out, reflective of the material in that chapter. The book can also be used as the basis of a course such as the one that I taught at New Jersey Institute of Technology (NJIT) and that

Professors Jablan and Radovic taught at their universities. It is an attempt to show how old and ancient ideas have had an influence on modern mathematics. I have attempted to present some mathematical principles in each chapter that capture the imagination, and each chapter contains mathematical surprises not widely known. Although its main purpose is to motivate the reader to carry out constructions based on the content, this is not merely a how-to-book. Rather, it is complete and consistent, supplying the background and history of the mathematical principles. Although the book is meant to be read by a wide audience, it does require some mathematical sophistication.

I now describe the content of the 26 chapters that comprise the book. Some years ago, Margit Echols, a great American quiltmaker, introduced me to a compass and straightedge construction known as the Vesica Pisces based on an equilateral triangle grid that opens up to a hexagonal pattern and is the key to many ancient designs. The *Vesica Pisces* is a pair of interlocking circles with a central fish-like region. It is an important Christian icon and can be seen on the facades of many European churches. Chris Hardaker, an archeologist, was successful in making this the key to introducing Native American fourth graders to mathematics with a course in geometry based on compass and straightedge constructions starting with the Vesica. Later it was pointed out to me that this construction was identical to a geometric construction known as the *Flower of Life*. This is the key to what is called sacred geometry from which groups of people from around the world derive particular beliefs based on unity and harmony. Since then, Margit Echols has introduced me to another ancient system based on a square grid also leading to wonderful designs and which Hardaker discovered in the proportions of Native American dwellings known as the Kivas of New Mexico. I felt that these constructions, based on triangular and square grids was the place to begin **Chap. 1** of this book.

Margit Echols showed me her Norman Conquest quilt pattern based on six fundamental patterns arranged in a square grid that I greatly admired. Each of the sub-patterns were derived from a simple construction known as the sacred cut in which the arc of a circle drawn from a vertex of a square through the square's center cuts the edge of the square in something referred to as the sacred cut. I had already known that the sacred cut was at the basis of much architecture from the Roman Empire as discovered by the architecture historians, Donald and Carole Watts. As Echols states, "In my search for new quilting ideas, I've always been attracted to the geometric designs that other craft traditions have explored. As long as I searched for new designs, I found few that looked worth the effort to piece. It was only when I started to visualize the obvious shapes in a design as composites of simpler shapes that I began to see new possibilities." It is this exploration that **Chap. 2** seeks to describe.

Chapter 3 begins with a set of Islamic designs based on the *Pythagorean theorem* going back to 900 AD. Some of these designs were discovered by the Islamic designer, W.K. Chorbachi, as part of her doctoral thesis from Harvard. Two of the proofs of the Pythagorean theorem are engraved on a wall mosaic in a mosque in Isfahan, Iran. Another proof leads to an algorithmic approach to creating designs which were most likely conveyed from master to artisan. We will present the proofs and indicate how to implement the designs.

In 1909, Z. Moron succeeded in tiling a rectangle with squares of different side lengths. Then there began a quest to find a tiling of a square by using only squares of different side lengths. It was not known if this could be done. Finally, it was accomplished in 1936 by four students from Trinity College. **Chapter 4** describes the less ambitious result in which a rectangle is tiled by non-congruent squares. After this was accomplished, it was noted that such tilings have analogues to electrical circuit diagrams.

In my design classes, my students created several examples of beautiful tilings based on wallpaper patterns in both 2 and 3-dimensions as shown in **Chap. 5**. Each tiling has a fundamental module that is a rectangle subdivided into numbered squares and rules by which these rectangles can proliferate through the entire tiling. In three dimensions, the module is a cube with numbered subcubes. Rotations of 90 deg. and 180 deg. are also incorporated in the rules. This introduces the reader to some of the elements of symmetry while creating lovely designs, some of which you can view on the book's website.

The Brunes star from **Chap. 6** is another discovery from the ancient world by the Danish engineer, Tons Brunes. I discovered his work in a two volume set of books entitled The Secrets of Ancient Geometry. Brunes felt that the star that he introduced and the sacred cut, whose expression he is responsible for, was the basis of ancient design. The star is easily created by placing the diagonals into two horizontal half squares and two vertical half squares which subdivides the space within the square into triangles and parts of triangles all related to 3,4,5-triangles. Brunes found his star pattern in tapestries in Pompei and Herculaneum. I found it in the home of Antonio Gaudi, the great Spanish architect in Barcelona, and also in the elevator shaft above the gift shop in the Sagrada Familia cathedral in Barcelona. When colored, this star leads to wonderful designs.

Slavik Jablan, who traced the history of design, felt that textiles, basketry, plaiting and weaving used these techniques in the most ancient times. He also discovered mosaics going back to 23,000 BCE in Mezin (Ukraine) to have used these tools. Many of these ancient designs used a set of parallel diagonal black and white stripes, or so-called *versatiles* which we will see again in the next chapter. There was no question that we should include in the book information as to how textile design, using these versatiles, dictate what a well-dressed Neolithic man should wear. And **Chap. 7** traces some of the history of these versatiles, asking the question, "Do you like Paleolithic Op-art?"

Chapter 8 goes into depth on versatiles. When turned at 90 deg. angles and juxtaposed with other tiles, you are able to create meandering linear patterns reminiscent of the circuit diagrams on computer chips. These versatiles can also be arranged to form spirals and labyrinths. The tiles can also be curved to create examples of *op-art*. **Chap. 8** also shows how a simple square, divided in half along the diagonal and colored black and white, can lead to an enormous diversity of designs, again, demonstrating how simplicity leads to complexity. These are called *Truchet tiles*. Finally, *Kufic tiles* are introduced in which a single black and white stripe is placed on a square and leads to Kufic writing found in Islamic mosques.

Meandering lines were introduced in the **Chaps. 7 and 8**. In **Chap. 9** the mathematics of *meanders* is studied using permutations of integers. It is interesting that Albert Einstein wrote a paper on meanders early in his career. These meanders have roots in the study of knots and links, labyrinths and mazes. My colleagues Slavik Jablan and Ljiljana Radovic applied their work to the study of a new class of meander knots and links. Kristof Fenyvesi, another colleague, wrote a short history of *Labyrinths* presented in Chap. 9. The Cretan Labyrinth of ancient Greek mythology is constructed along with a Labyrinth building workshop that took place at the Bridges conference, a conference on mathematics, architecture and design that has been held each year since 1998. These ideas also lead to the construction of *mazes.*

The *Mobius strip*, discussed in **Chap. 10**, is an excellent way to introduce children to what mathematics is about and is the theme of this book, namely, that the simplest of rules or objects are enough to bring you to unexpected places. Simply taking a strip of paper, giving it a twist and taping the ends and then cutting it down the middle is enough to create a result that is beyond your imagination. No further words are needed. If you give the strip a twist in the other direction and tape this Mobius strip at right angles to the first and then cut, it is amazing that you get a pair of interlocking heart shapes, something that the world needs more of. What a lovely result. This leads to the study of the properties of surfaces as the mathematical subject of topology, the basis of Chap. 10. Instructions to construct Mobius sculptures are presented along with information to construct a 4-dimensional surface known as the Klein bottle that has no inside and no outside. This also leads to a sculpture.

Among the concepts studied in topology is the *chromatic number* which is the greatest number of colors needed to color a map on a surface so that two faces sharing an edge have different colors. Since the 1970s, as proved by Appel and Haken, it is known that only four colors may be needed to color a map on the sphere or the plane. However, it was known long before, that sometimes seven colors are needed to color a *torus* or rubber tire. The Hungarian mathematician, Laslo Szilassi, was able to construct a torus-like polyhedron that had seven hexagonal faces with each face bordering on the other six. Another Hungarian mathematician, Akos Czaszar, constructed the polyhedron that was dual to the Szillassi polyhedron that has seven vertices with each vertex connected to the other six. In **Chap. 11**, Instructions and templates to build these polyhedra are given.

It is a great surprise in **Chap. 12** that you are able to construct non-circular wheels that roll just as well as the ones we are all familiar with. We can construct these so called *curves of constant width* beginning with an equilateral triangle and using compass and straightedge. Additional curves of constant width can be constructed with any set of intersecting line segments also using compass and straightedge. This construction also leads to a device that cuts square holes. We are also able to construct a 3-dimensional sculpture that rolls like a ball and is based on the geometry of an octahedron.

It has been only about fifty years since the notion of *fractal* swept upon the mathematical stage. Some mathematicians had come upon the concept of a fractal long before that by studying what were referred to as "pathological curves" while exploring the foundations of calculus. However, calculus only pertains to curves that are mostly smooth, while continuous curves were discovered that are nowhere smooth. In the 1970s, Benoit Mandelbrot, while studying the trends in the stock market, rediscovered these strange curves and coined the term *fractal*. Fractals model many images from nature such as lightening bolts, cauliflower stalks, and coastlines, and have many interesting properties such as being self-similar at many scales wherein any small portion of the whole fractal contains an exact replica. Furthermore, each fractal can be created by a set of rules applied over and over, ad infinitum. A cauliflower and meteorite markings on the moon are two such examples illustrated in **Chap. 13** along with the devilish Dragon curve.

Chapter 14 illustrates a procedure known as the Iterative Function System (IFS) in which rules of symmetry are applied to a black square over and over to create highly articulated fractal patterns. It clearly shows the self-similar nature of the pattern. Before generating the fractal, I will inform you about the rudiments of the symmetry of the plane and apply it to the symmetry of a square. We will also construct what I call the DNA of the fractal patterns using a "symmetry finder."

The tenor of this book changes in **Chaps. 15–21** in which, from the concept of a *logarithmic spiral*, the *golden and silver mean*s emerge with applications to architecture and design. **Chapter 15** shows how, by doing what I call "surgery on a right triangle," the logarithmic spiral can be derived as the end result of a geometric sequence. We explore Jay Hambridge's idea of *dynamic symmetry* which was known during the Renaissance as the *Law of Repetition of Ratios* through which a simple construction enables a rectangle to be replicated at different scales with a leftover portion called a *gnomon.* Dynamic symmetry leads to a set of *whirling gnomons* and a single proportional unit. In the process of applying dynamic symmetry, we come upon several proportions that are key elements of design, the golden and silver means, and the square roots of 2 and 3, each with their own design possibilities. Chapter 15 leads to the construction of a *Baravelle* spiral and spiral pathways, the result of four turtles chasing each other. For a golden rectangle, the gnomon is a square. This means that when you remove a square from a golden rectangle the result is a golden rectangle at a smaller scale. Now an excellent approximation to a logarithmic spiral can be constructed from a set of Whirling Squares.

This leads in **Chap. 16** to the golden mean, a number embodying self-similarity. The golden proportion or golden section is derived from a simple statement in which a line segment is broken into two parts such that the ratio of the whole to the largest part equals the ratio of the larger to the smaller. Self-similarity can also be represented by the *Fibonacci sequence* in which each term is the sum of the two previous terms. This leads to the golden mean as the ratio of successive terms from the Fibonacci sequence that approaches the golden mean closer and closer. The French architect Le Corbusier used a pair of Fibonacci sequences, the *Red and the Blue* to build a system of architectural proportions with additive properties that he called *Modulor.* If you succeed in tiling a square using rectangles from the Modulor system, then you are able to rearrange the tiles in numerous other ways to tile the identical square. If the system did not have additive properties, then you would never find another tiling beyond the first. This enables you to construct such a tiling using the Modulor of LeCorbusier, the subject of **Chap. 18.** The golden mean is also related to a regular pentagon which can be constructed using the property that the ratio of the diagonal to the edge of a pentagon is the golden mean.

What made me interested in the golden mean was *Wythoff's Nim,* a game played with pennies described in **Chap. 17**. Wythoff's Nim presents a winning strategy based on the golden mean revealing subtle properties of the golden mean leading to an infinite sequence of 1's and 0's with applications to plant growth and dynamical systems.

In 1978, Roger Penrose surprised the world by creating a pair of *kite and dart* shapes that enabled what is referred to as *nonperiodic tilings* of the plane to be constructed. This enabled tilings constructed with approximate five-fold symmetry, whereas the standard theory of symmetry denies this possibility. It also resulted in pleasing designs as **Chap. 19** illustrates. It was recently discovered that Islamic *Girih* tilings appear to be related to the Penrose tilings yet were discovered in the 13th century. It is unlikely that the Islamic mathematicians were thinking in the 13th century about kites and darts, but it is nevertheless fascinating that they appear to have stumbled on them.

In **Chap. 20**, we come upon another important number called the silver mean that was derived from the sacred cut and led to the construction of an octagon. We have previously mentioned that Margit Echols used the sacred cut to create her quilting patterns. The silver mean is a number that also has many additive properties and leads to another system of proportions that was employed during the time of the Roman Empire as discovered by Donald and Carole Watts in their study of the Garden Houses of Ostia, a major port city of the Roman Empire. Just as the golden mean was derived from the Fibonacci sequence, the silver mean is the result of the Pell sequence in which each term is the sum of twice the last term and the one before that. Just as the Modulor of Le Corbusier was derived from a pair of numbers from the Red and Blues sequence related to a six-foot British policeman with his hand held over his head, the silver mean is expressed by an infinite sequence of Pell sequences, the numbers of which can be derived from the numbers 1 and $\sqrt{2}$ and therefore can be constructed with compass and straightedge.

I learned that natural systems of proportion are rare. In fact, I believed that the golden and silver means are the only systems that truly work. The numbers within these systems are placed so that they relate to their neighbors as geometric, arithmetic and harmonic means. It was a great surprise when I learned that the Roman mathematician from the second century AD, Nicomachus, created sequences of numbers from which the pentatonic, heptatonic and chromatic musical scales can be expressed. It was these sequences of numbers that led the Renaissance architect Alberti to create another system of proportions based on the musical scale with identical structure to the systems of Roman architecture and the Modulor in terms of their *arithmetic, geometric, and harmonic means.* **Chapter 21** presents what I call a unified system of proportions to illustrate these relationships. Since the musical scale is at the base of this system, I have shown how the musical scale can be derived from the integers 2 and 3. During the time of ancient Greece, 2 was the first female number while 3 was the first male number. I show that there was a logic to these designations and show how to derive musical scales from the numbers 2 and 3.

Chapter 22 introduces the concept of a *zonogon* which is a polygon with opposite sides parallel and equal. The edges of each polygon are defined from a star of vectors, referred to as an *n-vector star.* The vectors from the star can be lengthened and shortened to suit your interests, and during these transformations the angles do not change. For example, you can tile the plane with 3-zonogons that represent hexagons with odd shapes.

Tangram is a parlor game invented in the early 1800s. The tangram set has seven squares and half-squares and one tangram diamond that you get from putting two half squares together. The half-squares come at three different scales. From these pieces you are able to assemble an almost unlimited set of pictograms using each of the seven pieces with no repeats as described in **Chap. 23**. Around the time that I was playing with tangrams, I paid a visit with my family to Amish country around Lancaster, Pennsylvania looking for things that I could do with my students when school began. I purchased a quilt with parts that looked very much like the tangram pieces. However, instead of the tangram diamond, Amish quilts are often constructed with rhombuses having angles the same as tangram diamonds but with all sides equal. This, I refer to as an *"Amish diamond"*. Also, the diagonals of the Amish diamond turned out to cut each other in the sacred cut. So, this was a great opportunity to have my students create Amish quilt-like patterns using the tangram pieces and Amish diamonds.

It turns out that two squares and four Amish diamonds combine to create a regular octagon. It occurred to me that the Amish diamond is an example of a zonogon. It turns out that n-zonogons with all vectors the same length are planar projections of polyhedra referred to as *zonohedra, an area* explored by *Professor of Architecture, Haresh Lalvani.* Very nice designs can be made from 4, 5 and 6-zonogons. The 4-zonogon is the planar projection of a polyhedron called the rhombic dodecahedron. A 5-zonogon results in patterns involving the golden mean, and 6-zonogons are projections of a polyhedron called a *rhombic triacontahedron* which is related to a 6-dimensional cube. These themes in Chap. 23 were food for thought in creating the material for my design class.

Slavik Jablan has introduced a construction in which a light ray runs through a matrix of dots, bouncing off one-way mirrors placed, along the edges of the matrix and surrounding the dots. Internal two-way mirrors can also be added in the creation of these *mirror curves*. Sometimes all the dots can be surrounded in a single stroke, but it can also take several strokes. This set of circumstances was turned into a game played by the natives of the Tchokwe tribe of Angola and studied by Paulus Gerdes, an ethnomathematician from Mozambique. **Chapter 24** leads you to create these mirror curves. Then in **Chap. 25,** by following the light rays on a single curve traversing the matrix, highly structured black and white patterns called *Lunda designs* are created as described by Prof. Radovic. Finally, a set of "orn" tilings are created that wind their way through a square lattice.

From the work of Jablan and Radovic, **Chap. 26** leads to a study of visual illusions describing constructions that can be made into op-art designs and impossible figures, some of which can be found in the work of M.C. Escher.

It is forty years since I became engaged with the topic of mathematics and design. It is gratifying to see how others have brought it into a self-contained discipline. There is now a monthly journal, Mathematics in the Arts. Many of the ideas have filtered down to pre-college education. Several conferences on this subject are held each year. Through the 1990s, I attended a conference on math and art held each year at SUNY Albany, organized by Nat Friedman, a mathematician and sculptor. I have been a regular participant at the Bridges Conference which originated in 1998 by Reza Sarhangi and the Symmetry Festival held every three years under the leadership of Gyorgy Darvas. There is a movement in academia to encourage interdisciplinary learning, and this is ideally suited to a study of the relationship of math and design. Tools have been created to engage the computer in a meaningful way through advanced computer graphics programs such as Adobe Write, Corell Draw, Photo Shop and Sketchup. Along with the advent of 3D-printing, this points towards a bright future for the new disciple of Mathematics and Design.

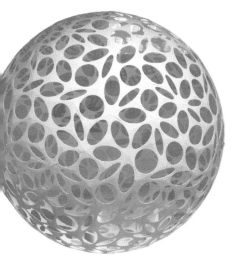

Acknowledgements

I wish to acknowledge my two colleagues, Slavik Jablan and Ljiljana Radovic who have provided the inspiration and many of the ideas that have been essential to the creation of this book. In fact, it was Slavik who fostered the idea that we should collaborate on the writing of this book before his untimely death. He and Prof. Radovic provided several of the chapters you will find in this book. I also thank Haresh Lalvani whose lovely spherical sculpture adorns the title page of each chapter and the back cover of the book. Several other of his sculptures can be found in Chapter 23 along with some of his ideas about zonogons featured in that chapter. My thanks goes out to Werner Van Hoeydonc who contributed his inspired design to adorn the front cover of this book.

I also wish to acknowledge my former student, Kevin Miranda, whose graphic skills adorn the pages of the book. He also helped test some of the projects found in this book. I also acknowledge my colleague Denis Blackmore who read the book and made important comments, and Dick Esterle who made helpful remarks throughout the writing of this book. He also created a 3-dimensional sculpture related to an octahedron that rolls found in Chap. 12. Judith Friedman used her editing skills to assure that the book meets high standards of English expression. I also wish to thank my editor, Tan Rok Ting, at World Scientific, whose patience gave me the space to

xxvii

complete this complex book. I also wish to thank Soh Jing Wen, my other editor, who meticulously reviewed, with great care, every word, every sentence, and every paragraph.

Finally, and most important were the generations of students who used their talent and skills to create the wonderful designs that you will find throughout the book and on the website.

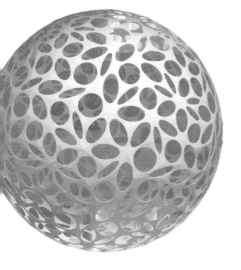

Chapter 1

Triangle-circle and Square-circle Grids

1.1. Introduction

It is amazing the power and potential that exists by the simple act of drawing a circle or a square. This was recognized in ancient times where the circle symbolized the celestial sphere and represented the Heavenly domain; the square, with its reference to the four directions of the compass, represented the Earthly domain. Margit Echols [Ech] was a quilter who used geometry as the basis of her wonderful quilt patterns. In her workshop Echols shows how to create triangle-circle and square-circle grids beginning with a point, line and circle which becomes scaffolding for the creation of a limitless number of repeating patterns. These patterns are sometimes referred to by sacred geometers as The Flower of Life [Flo]. You can see many other designs on the book's website.

1.2. Triangle-Circle and Triangle Grids

For this exercise, you may use either a compass or a software program such as Corel Draw, Adobe Illustrator, or Geometer's Sketchpad. To create a triangle grid, start with a circle. Place the compass point on the circumference and draw another circle with the same radius. The resulting pair of circles is perhaps the most fundamental structure in geometry called the *Vesica Pisces* shown in Fig. 1.1a. This figure

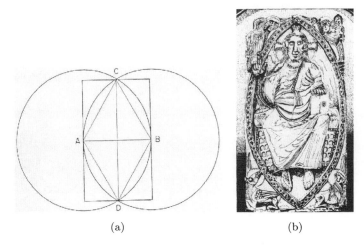

(a)   (b)

Fig. 1.1.   (a) The Vesica Pisces (b) Marble relief of Christ in vesical

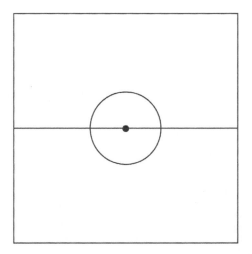

Fig. 1.2.   Geometry begins with a point, a line and a circle

had sacred significance in the Christian religion. It is found in many churches where images of Christ were drawn in the central fish-like region as shown in Fig. 1.1b. The Vesica Pisces is created by starting with a line and a single point on the line which is the center of a circle as shown in Fig.1.2. When the line intersects the circle, draw another circle of the same radius. The *Vesica Pisces* is the starting

Drawing Circle and Triangle Grids

1. Draw a circle.

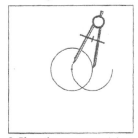

2. Place the compass point on the circumference and draw another circle.

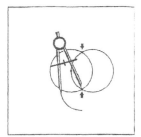

3. Place the compass point at points of intersection and draw 2 more circles.

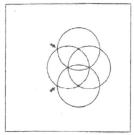

4. Place the compass point at new points of intersection and draw 2 more circles.

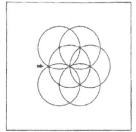

5. Place the compass point at the last intersection on the first circle and draw a circle.

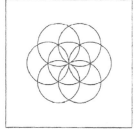

6. The circles divide the first circle into 6 equal parts.

7. To make a circle grid, continue adding circles until the page is filled.

8. Connect the centers of the circles to create a grid of equilateral triangles.

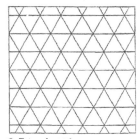

9. Triangle grid.

Fig. 1.3.   Drawing a triangle-circle grid

point of a *triangle-circle* grid. Figure 1.3 shows how to generate such a grid step by step. By step 9 you have created a triangle grid. Look within the grid. Can you see a grid of hexagons embedded in the grid of equilateral triangles?

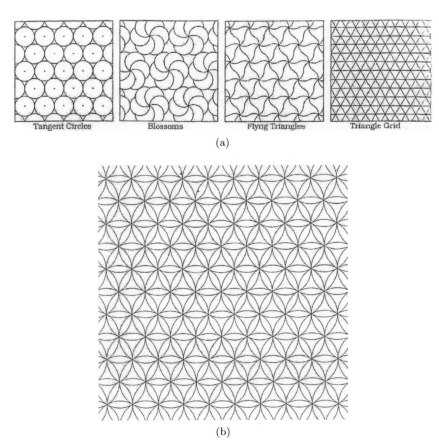

Fig. 1.4.   (a) Pattern puzzle A; (b) Overlapping circles: triangular system

Pattern Puzzle A is presented in Fig. 1.4a. These patterns are hidden in the triangle-circle grid of Fig. 1.4b. Draw each of these patterns using the overlapping circles in Fig. 1.4b as a guide. You can add color. Note that some patterns can be traced directly from the curved lines, while the triangle grid can be drawn using a straightedge to connect points where the circles intersect. Many other patterns can be found in this triangle-circle grid. Try to find others, or invent some of your own. The possibilities are endless!

Pattern Puzzle B is presented in Fig. 1.5a. Look for these patterns hidden in the triangle grid in Fig. 1.5b. You can add shading or color. You can also draw each one on a sheet of tracing paper

PATTERN PUZZLE B

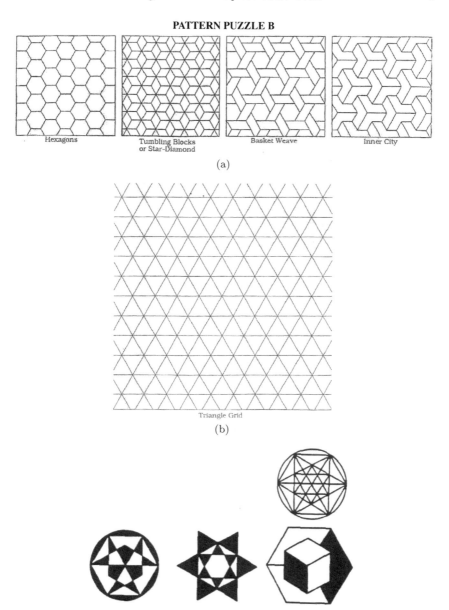

Hexagons

Tumbling Blocks
or Star-Diamond

Basket Weave

Inner City

(a)

Triangle Grid

(b)

The versatility of the triangular grid. These
are created by shading a portion of the grid enclosed by
the circle.

(c)

Fig. 1.5. (a) Pattern puzzle B; (b) Triangle grid; (c) Versatility of the triangular
grid. Can you reproduce these patterns from the grid enclosed by the circle?

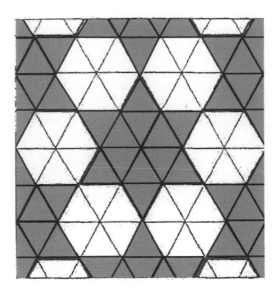

Fig. 1.6.   Islamic design with hexagons and triangles

using the underlying grid as a guide or use your computer drawing programs. Four additional patterns are shown in Fig. 1.5c. You may try your hand at creating patterns of your own. Again, there are an infinite number of possibilities. Islamic artists designed patterns with hexagons and 6-pointed stars on this grid from the center outward with no gaps and no overlaps as can be seen in the shaded stars of the triangular grid in Fig. 1.6.

1.3. Square-Circle and Square Grids

Whereas the triangle-circle grids were created by beginning with a line and a single point on it, square-circle patterns begin with a pair of lines intersecting at a right angle. The point of intersection of these lines defines the center of a circle. A circle is drawn about this center in order to define the four points where the pair of perpendicular lines intersect the circle. The initial circle is then removed. The four new points are centers of new circles with the same radius. The points

Drawing Circle and Square Grids, and Star-Cross Pattern

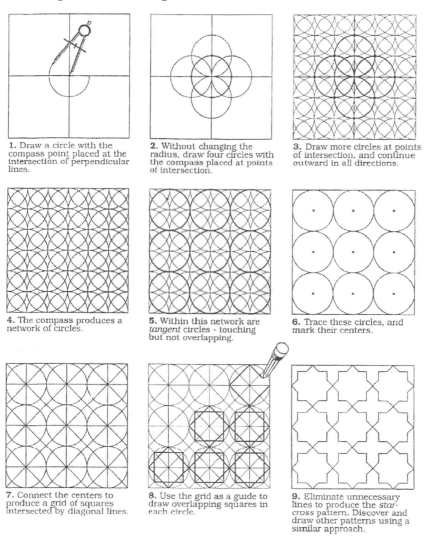

1. Draw a circle with the compass point placed at the intersection of perpendicular lines.

2. Without changing the radius, draw four circles with the compass placed at points of intersection.

3. Draw more circles at points of intersection, and continue outward in all directions.

4. The compass produces a network of circles.

5. Within this network are *tangent* circles - touching but not overlapping.

6. Trace these circles, and mark their centers.

7. Connect the centers to produce a grid of squares intersected by diagonal lines.

8. Use the grid as a guide to draw overlapping squares in each circle.

9. Eliminate unnecessary lines to produce the *star-cross* pattern. Discover and draw other patterns using a similar approach.

Fig. 1.7.   Drawing a square-circle grid and the star-cross pattern

where two circles intersect define the centers of new circles of the square-circle grid. Construct a square-circle grid by following the step by step procedure in Fig. 1.7. This grid again leads to countless designs, 16 of which are shown in Fig. 1.8.

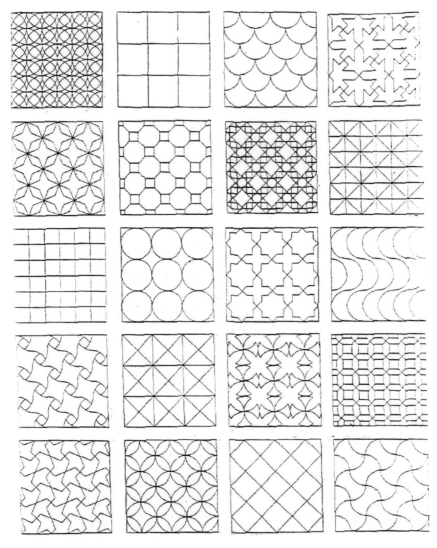

Fig. 1.8.   Sixteen examples of square and square-circle grids

1.4. Star Designs Based on the Triangle and Square Grids

Star patterns are powerful geometric objects and play a large role in this book. Every religion has adopted the iconology of a star through which it is recognized: 16-pointed stars for the Buddhist religion,

Drawing Six- and Twelve-Pointed Stars

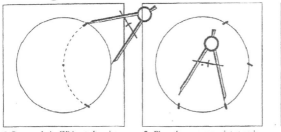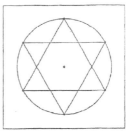

1. Draw a circle. Without changing the radius, place the compass point on the circumference, swing it to each side, and make two marks. These marks *intersect* the circle.

2. Place the compass point at each intersection, swing the compass to each side as before and make two more marks. Repeat until the circle is divided into six equal parts.

3. Connect every other point to make a six-pointed star. Note the hexagon in the center. How many equilateral triangles does the star contain?

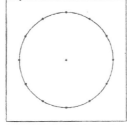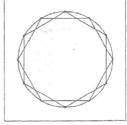

4. Draw lines through the center of the star and opposite vertices of the hexagon...

5. ...to divide the circle into twelve equal parts.

6. Connect every second point to produce two overlapping hexagons.

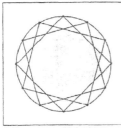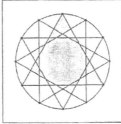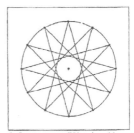

7. Connect every third point to produce three overlapping squares.

8. Connect every fourth point to produce four overlapping triangles.

9. Connect every fifth point without lifting the pencil from the paper to produce a "true stellation."

Fig. 1.9. Drawing 6 and 12 pointed stars

6-pointed stars for the Jewish and Seventh day Adventists, and 8-pointed stars for Christian and Islamic religions.

Beginning with a compass and straightedge construction as shown in Fig. 1.9, 6 and 12-pointed stars can be constructed. For this exercise you can also use a coffee can cover instead of a compass

Stars

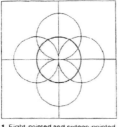

1. Eight-pointed and sixteen-pointed stars are related to the square grid construction. (See previous panel.) Connect the center...

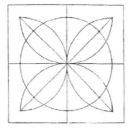

2. ...with points at which the four outer circles intersect.

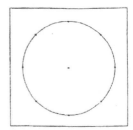

3. This divides the circle in eight equal parts.

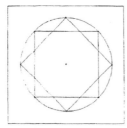

4. Connect every second point to produce an eight-pointed star with overlapping squares and an octagon in the center.

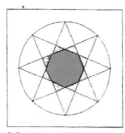

5. Connect every third point without lifting pencil from paper to create a "true stellation." Note that crossing lines produce the 8-pointed star of Fig. 4.

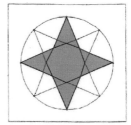

6. A popular four-pointed star is embedded in the eight-pointed star.

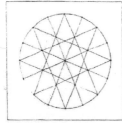

7. Connect opposite points in the central eight-pointed star to divide circle into sixteen equal parts.

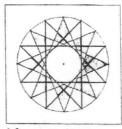

8. Connect every sixth point to produce a sixteen-pointed star with overlapping eight-pointed stars.

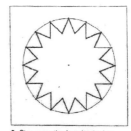

9. Stars were the favorite design element of Islamic artists.

Fig. 1.10. Drawing 4, 8, and 16 pointed stars

to carry out the construction in which the center of the cover has the usual marking; you simply place the center of the coffee can cover over the point that you would otherwise place your compass point to draw a circle. Beginning with the square-circle grid, 8 and 16-pointed stars can be created as shown in Fig. 1.10.

Construction 1: Use Adobe Illustrator, Corel Draw or simply a compass and straightedge to create a triangle-circle grid and a triangle grid beginning with a circle. After doing this, create your own pattern based on the triangle-circle grid and the triangle grid. You will find several examples on the website.

Construction 2: Beginning with a circle, repeat the instructions for Construction 1 to construct a pattern of your own based on square-circle and square grids.

Construction 3: Repeat the instructions for Construction 1 and create star patterns based on 6, 8, 12, and 16 pointed stars. Alternatively, you can experiment with other stars of your choosing. You may enlarge the circle broken into 360 parts in Fig. 1.11. Be bold and try to create an outrageous star. See examples on the website.

Fig. 1.11.   Circle divided into 360 parts

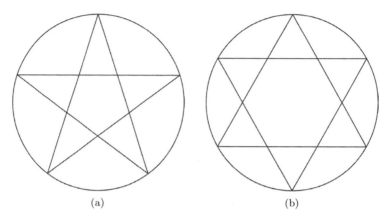

Fig. 1.12.   (a) $\{5, 2\}$ star; (b) $\{6, 2\}$ star

1.5. Star Exploration

Consider the two stars shown in Figs. 1.12a and 1.12b. The first star (see Fig. 1.12a) has 5 points in which every second point is connected. I call this a $\{5, 2\}$ star. The star in Fig. 1.12b has 6 points in which every second point is connected, i.e., $\{6, 2\}$. Do you notice that there is something strikingly different about these two stars? The $\{5, 2\}$ star can be drawn in one stroke without taking your pencil off of the paper, while the $\{6, 2\}$ requires the star to be drawn in two strokes.

Exploration: Draw a number of $\{n, p\}$ stars, i.e., stars with n-points evenly distributed around a circle in which every p-th point is connected. Try to come up with a condition on n and p that enables you to predict whether a $\{n, p\}$ star can or cannot be drawn in a single stroke and how many times you have to take your pencil off of the paper to draw the star. You can use the circle with 360 divisions it in Fig. 1.11 to help you in your exploration. Enlarge this star and replicate it.

1.6. The Two Great Systems of Ancient Geometry

The triangle-circle and square-circle patterns can be thought of as coming from two universes of ancient design both constructible with compass and straightedge.

a. Triangle-circle patterns lead to equilateral triangles. Dropping an altitude from a vertex to the base of such a triangle leads to a 30, 60, 90 deg. triangle.

b. Square-circle patterns lead to squares. Dividing a square by its diagonal leads to a 45, 45, 90 deg. triangle.

These two systems were of great importance in the construction of the architectural masterpieces of antiquity, and they will play a large role in the remainder of this book.

Five and ten-pointed stars are also of great importance and will be introduced in Chap. 16.

1.7. Regular and Semiregular Tiling the Plane

The tiling of the plane with equilateral triangles, as shown in Fig.1.5b, is an example of a *regular tiling* of the plane since each vertex is surrounded identically by a single regular polygon. For example, six equilateral triangles surround each vertex of the tiling. This tiling is coded by the *Schlafli symbol* $\{3,6\}$ which is self-explanatory. If we place a vertex in the centroid of each triangle and connect the vertices, we get the dual tiling of three regular hexagons surrounding each vertex, or $\{6,3\}$. Next, consider a tiling with squares as in Fig. 1.7 where four squares surround each vertex, i.e., the Schlafli symbol is $\{4,4\}$. This tiling is self-dual. The three regular tilings of the plane are shown in Fig 1.13a.

Now that we have identified the three regular tilings of the plane, let us relax the condition that the tiling should have a single regular polygon but still require each vertex to be surrounded identically by more than one regular polygon, i.e., the sum of the vertex angles of these, so called, *semiregular tilings* is 360 deg. They are also referred to as *Archimedean* tilings. There are 8 additional semiregular tilings as shown in Fig. 1.13b. One such tiling is shown in Fig. 1.14a with a second Schlafli symbol $3^2.4.3.4$, since you encounter a sequence of triangle, triangle, square, triangle, square around each vertex. Figure 1.14b shows that its dual is a tiling by congruent pentagons (not regular). It follows that all tiles of the semiregular duals are congruent.

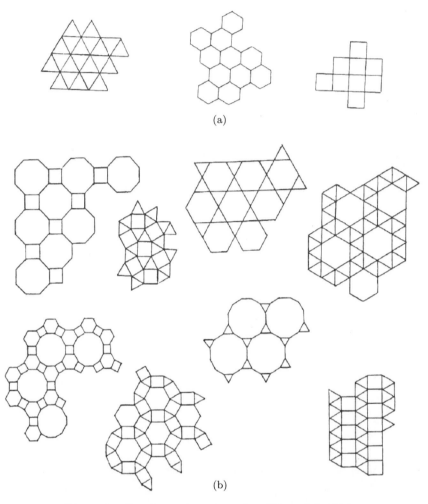

Fig. 1.13. Regular and semi-regular tilings of the plane

There is one point of caution here. It is not enough that three triangles and two squares surround each vertex in $3^2.4.3.4$. They must surround the vertices in the same order. For example, Fig. 1.15 represents two tilings with 2 squares and 3 triangles. Without this restriction, there can be an unlimited number of such tilings.

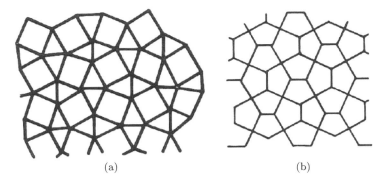

Fig. 1.14.   (a) and (b) — The Archimedean tiling $3^2.4.3.4$ and its dual

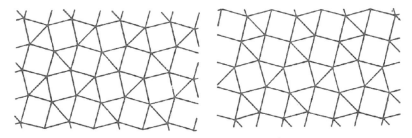

Fig. 1.15.   Two tilings with the same the same polygons but with different orderings

1.8.   The Module of a Semiregular Tiling

A manufacturer wishing to produce a set of tiles that cover the plane in a semiregular fashion does not have to create all the tiles individually. Each tiling has a basic *module* which can be rigidly moved to stamp out the entire tiling. Let's determine this module for a typical tiling, 3.6.3.6. Several elements of this tiling are shown in Fig. 1.16a along with its dual tiling. As you can see, a typical tile of the dual is made up of 1/6 of each of two of the original's hexagons and 1/3 of each of the original's triangles. Thus, since all tiles of the dual are congruent, the tiling must have hexagons and triangles in the ratio, 2/6 hexagon: 2/3 triangle or 1 hexagon: 2 triangles.

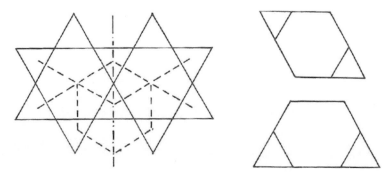

Fig. 1.16.   (a) — A portion of 3.6.3.6 with its dual superimposed; (b) — two modules of the 3.6.3.6 tiling

Figure 1.16b shows two such modules. You can see that this module can be translated to generate the entire 3.6.3.6 tiling.

1.9.   The Krotenheerdt Tilings

Whereas there are 3 regular tilings of the plane and 8 semiregular tilings, O. Krotenheerdt discovered that there are exactly 135 n-uniform tilings where n takes values no greater than 7 [Gru]. A 7-uniform tiling is shown in Fig. 1.17. In Krotenheerdt tilings, all

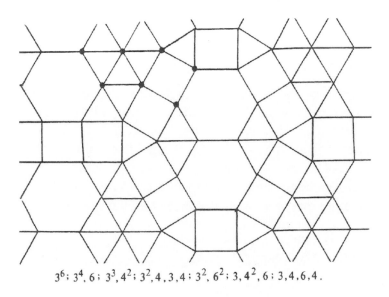

3^6; 3^4, 6; 3^3, 4^2; 3^2, 4, 3, 4; 3^2, 6^2; 3, 4^2, 6; 3, 4, 6, 4.

Fig. 1.17.   A 7-uniform tiling

Table 1.1. The number of n-uniform
Krotenheerdt tilings.

n	1	2	3	4	5	6	7
N	11	20	39	33	15	10	7

polygons are regular while there are several different classes of vertex that are surrounded identically. For example in Fig. 1.17 there are exactly 7 classes of vertices listed with their Schlafli symbols of the second type. The 7 darkened points represent the 7 Krotenheerdt centers within the tiling. Table 1.1 lists the number of Krotenheerdt tilings for a given value of n. I list several Schlafli symbols for different Krotenheerdt tilings in Construction 2. Additional Krotenheerdt tilings can be found on the website.

Constructions:

1. Construct the duals for several of the semiregular tilings. Try superimposing their duals over the tilings
2. Try your hand at constructing several Krotenheerdt tilings from the following list of Schlafli number for the given, n-uniform tilings

$$2\text{-uniform:} \quad 3.4.6.4. \; ; \; 3^3 4^2$$
$$3^6 \; ; \; 3^2 4.3.4$$
$$3\text{-uniform:} \quad 3^3 4^2 \; ; \; 3^2 4.12 \; ; \; 3.4.6.4$$
$$3^6 \; ; \; 3^2 4.3.4 \; ; \; 3.4.6.4$$
$$4\text{-uniform:} \quad 3^6 3^4 6 \; ; \; 3.4.4.6 \; ; \; 3.6.3.6 \; ; \; 4^4$$

1.10. Polygons that Fill Space by Themselves

Conway [Con] has written about the condition by which a single polygon, including curvilinear polygons, can fill space. I consider here the following polygons that fill space by themselves:

a. All triangles
b. All four-sided polygons (quadrilaterals)
c. All hexagons with opposite sides parallel and equal

d. Regular pentagons do not fill space by themselves. However, I invite you to explore the patterns that they make in the attempt to fill space.

Construction: Give examples of these space-filling polygons that fill space by themselves.

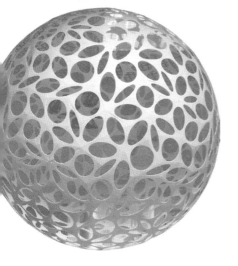

Chapter 2

Margit Echols' Magic Squares

Margit Echols was an American quilter who used geometry to create her quilts [Ech]. How did she use the square grid to create the beautiful "Norman Conquest" pattern illustrated in Fig. 2.1? Within the quilt pattern you can find the six pictograms shown in Fig. 2.2. The quilt pattern is entirely based on a square grid and a "magic square" whose construction is shown in the four-step process of Fig. 2.3 resulting in the "Star Cross" pattern. This construction will be shown to have deep significance in Chap. 20 where it is derived from a construction known as the *sacred cut*. The sacred cut is a subdivision of the edge of a square gotten with compass and straightedge construction by placing the compass point on a vertex of the square and drawing an arc through the center of the square until it intersects an edge of the square, as shown in Fig. 2.3. The side of the square is then said to be sectioned by the sacred cut. By step 3, eight points appear on the edges of the square. In step 4 the dots are connected by four lines to form the Star Cross pattern.

In Fig. 2.4a, this method is applied to connecting the dots on the edge of the Magic Square to recreate the pictograms of Fig. 2.2 that comprise the Norman Conquest. This is step 1 of a three step-process. First the square is divided as in Fig. 2.4a; these divisions of a square are then placed in a square grid as shown in Fig. 2.4b; the square grid is then removed to reveal the final pattern as shown in Fig. 2.4c.

Fig. 2.1. The Norman Conquest quilt

Listen to what Echols says about this approach to quilt pattern making:

"The key to creating the Star Cross pattern is to recognize that the star is based on a square, as shown in Fig. 2.3, and that once you have the square pattern, the entire pattern is based on a square grid, as shown in the diagram(s) of Fig. 2.4. You can see how this same logic will unlock many other patterns some of which I used in my Norman Conquest quilt. In fact, all of the patterns in Norman Conquest can be derived from the same square grid and can be broken down for piecing in similar ways.

Design books are filled with patterns in which you can find an underlying — sometimes obvious, sometime obscure — square grid, and from this you can develop a simple piecing solution. Of course,

PIECING
CHALLENGES

*Cut without internal
sears, none: of these
shapes would be easy
to picas.*

Fig. 2.2.   Six pictograms in the quilt pattern

not all patterns are based on a square grid. If you discover another type, such as a triangle grid, try using the same approach.

An easy way to explore these designs is to make photocopies of the pattern and experiment on them with a pencil and a ruler to find the underlying grid. Once you find it, you'll notice that it is usually possible to simplify it just by adding a few lines within the square subunit, or by changing a curve to a straight line or vice versa.

And what's behind the square grid itself that makes it such an inexhaustible source of patterns you can design yourself and which dates back at least as far as ancient Egypt? But that is another story."

I have presented Echols' approach to pattern creating. I leave the details of her piecing technique to her many books. Margit Echols

The Magic Square

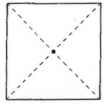
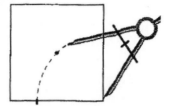

1. Draw a square, any size, on graph paper. Find the center by lightly drawing two diagonal lines connecting opposite corners.

2. Place the compass point at one of the corners and adjust the radius until the lead meets the center of the square. Swing the compass right and left, and mark where arc intersects the adjacent sides of the square.

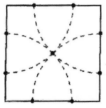
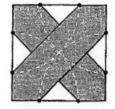

3. Repeat at each corner to produce a four-leaf clover design. Note the eight points where the arcs cross the square.

4. Connect these points as shown to make the cross block for Star Cross. Connect the points in differentways to make different designs.

Fig. 2.3.　Using the magic squares to draw the Norman Conquest pictograms in a square lattice. (a) dividing the square, (b) the squares are revealed, (c) final pattern

has written 50 books on quilting. They are available on Rodale Press. Her approach to piecing can be found in Threads No. 55 pp. 56–59.

Exercise 1. Try using the method shown in Fig. 2.3 to recreate the final patterns in Fig. 2.4. Try to find a pattern of your own using the techniques of this chapter.

Exercise 2. If you are a quilter, try to replicate the Norman Conquest or make your own quilt.

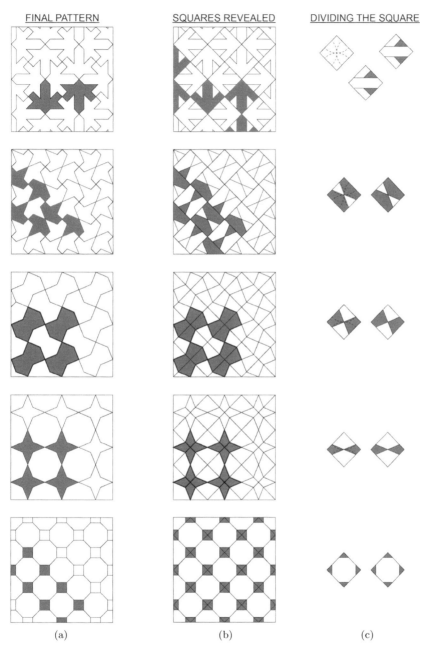

Fig. 2.4. Genesis of the patterns within the Norman Conquest quilt. (a) Dividing the square, (b) The square grid is revealed, (c) The square grid is removed revealing the final pattern

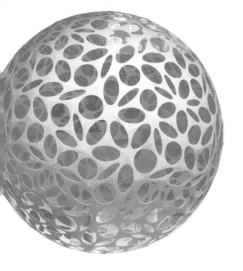

Chapter 3

The Pythagorean Theorem as a Theme in Islamic Art: An Algorithmic Approach

3.1. Introduction

W.K. Chorbachi, an Islamic artist and designer, discovered while carrying out the research for her doctorate at Harvard the increasing use of complex geometry in the creation of Islamic art and decoration beginning around 900 AD [Cho]. Her findings seemed to contradict the thinking prevalent at the time that Islamic artisans could not have been sophisticated enough to carry out such geometry. She records in her article, written in 1989 in *Computers in Mathematics with Applications*, the story of how she discovered the primary documentation that reveals the connections of advanced geometry underlying Islamic design. She did this by showing how Islamic geometry masters communicated information to the artisans who then executed the designs. The masters insisted on logical rigor and would not accept an approximate rendering of a design. The design must follow a step-by-step re-creation, generally using compass and straightedge constructions. Contrary to warnings from her professors at Harvard that nothing would be found, she scored a success only two weeks after beginning her search and actually found a document entitled "A treatise on what the artisan needs of geometric

25

problems", written by one of the luminaries of Islamic mathematics from the 13th century, Kamal al-Din Yunis bin Man'a. This chapter will focus on Islamic designs motivated by proofs of the Pythagorean theorem.

I should also mention that in doing her research, Chorbachi looked for and found the language of symmetry to be invaluable and sought the help of Arthur Loeb, a chemical physicist, who often came to the aid of artists and designers needing an understanding of modern mathematics in their work. The spirit of Chorbachi's work is closely in sync with the following statement made by Arthur Loeb [Loe 1] in his book, Algorithms, Structures and Models:

> "We have observed that our perception of an apparently complex configuration is altered when, instead of attempting a complete description of the object, we generate the configuration from a small number of relatively simple modules together with an algorithm for assembling them.
>
> Generally, we do not know the modules and algorithms that would generate a given complex configuration. The role and process of science would seem to consist of a search for appropriate modules and algorithms to generate models whose behavior resembles adequately to that of the complex configuration studied. The analogy between the model and observed configuration is limited and quite subjective, depending on the observer and the purpose, context and background of the experiment.
>
> In design, the algorithmic approach generates, with simple means, a rich repertoire of patterns transcending the repertoire of the 'naked eye'. In addition, the conceptual component of such a generated pattern has an aesthetic appeal of its own and constitutes an important link between art and science."

The purpose of this chapter is to give an example of how this algorithmic approach might have been implemented for designs motivated by the Pythagorean theorem. In applying this to Islamic design, Chorbachi found that art must search for the proper scientific languages and tools to generate new forms and expressions. In this chapter, the Pythagorean theorem represents one such starting point for the genesis of designs.

3.2. The Pythagorean Theorem

I begin by describing four proofs of the Pythagorean Theorem which have roots in Islamic design and which bear on Chorbachi's discoveries. I then proceed with a short geometry lesson needed to understand the designs that Chorbachi discovered. A step-by-step rendering of one ingenious construction follows, leading to an algorithmic approach to the construction of countless designs, a path the artisans might have taken.

The Pythagorean theorem is perhaps the most important theorem in all of Euclidean geometry. No one knows its origin, however, it was probably known long before the age of Pythagoras (570–495 BCE). There have been more than seven hundred proofs of this great theorem. In this chapter we will describe four of them and show how the Pythagorean theorem may have inspired several Islamic designs. After describing the Pythagorean content of an iconic design from Isfahan, circa 1700, we will then focus on a discovery that W.K. Chorbachi found in a set of notebooks entitled Folio 192b housed in the Bibliotheque Nationale in Paris [Blochet, 1912]. This proof was popularized by the puzzle master, Martin Gardner [1978].

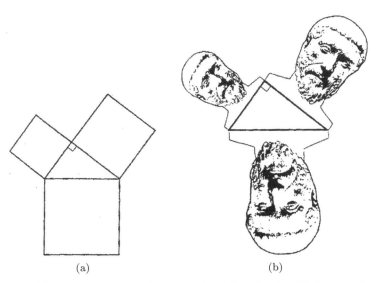

(a)   (b)

Fig. 3.1.   (a) Three squares on the edges of a right triangle; (b) three pythagoras on the sides of a right triangle

The Pythagorean theorem states that if you place a square on each of the sides of a right triangle, the area of the square on the hypotenuse equals the sum of the areas of the two squares on the sides of the triangle (see Fig. 3.1a). For that matter, as Fig. 3.1b shows, you can just as well place Pythagoras on each edge of the triangle; the area of the Pythagoras on the hypotenuse then equals the sum of the areas of the Pythagorases on the other two sides.

3.3. Four Proofs of the Pythagorean Theorem

3.3.1. Pythagorean theorem by dissection of a square

This proof can be attributed to Bhaskara, an Indian mathematician of the 12th century AD. In Fig. 3.2a you see a right triangle with area A_T and sides a, b, c, and beside it, in Fig. 3.2b, an outer large square with area A_L divided into 5 areas, four triangles of total area $4A_T$ and a smaller interior square of area A_S.

Expressing the areas of A_T, A_L, A_S in terms of a, b, c and replacing them in:

$$A_L = A_S + 4A_T$$

states that the sum of the 5 areas equals the area of the outer square.

After some algebra in which A_L, A_S, A_T are expressed in terms of the edge lengths a, b, c, results in,

$$c^2 = a^2 + b^2,$$

a proof of the Pythagorean theorem for the original right triangle.

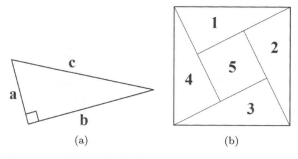

(a)   (b)

Fig. 3.2.   (a) Proof of the Pythagorean theorem by dissection of a square

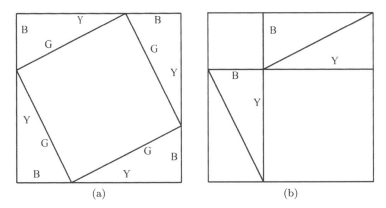

Fig. 3.3.   An *Aha!* proof of the pythagorean

3.3.2.   An "*Aha!*" proof

a. This proof of the Pythagorean theorem is an "*Aha!*" proof in which the Pythagorean theorem can be observed without computation. This means that the validity of the theorem should be evident by just looking at the diagram. In Fig. 3.3a. four congruent right triangles have been arranged in the outer square leaving the inner square empty.

   Notice that this square lies on the hypotenuses of the right triangles each of which shares a green edge with the right triangles, and where the edges of the triangles are color coded with the colors blue, green and yellow.

b. In Fig. 3.3b the triangles have been rearranged leaving two square areas empty with each square lying on a blue and yellow edge of the right triangle. Notice that the sum of the areas of these squares equals the area of the original square so that the square on the hypotenuse equals the sum of the squares of the edges of a right tringle, a statement of the Pythagorean theorem as: $G^2 = B^2 + Y^2$.

3.3.3.   Pythagorean cake cuts

The following proof was reported by Martin Gardner [Gar1]. In Fig. 3.4a, a diagram of a cake with two equal perpendicular slices has been drawn, resulting in four congruent quadrilaterals.

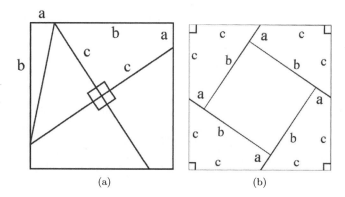

Fig. 3.4.   Proof of Pythagorean theorem by Gardner's cake cuts

The sides of each cake slice are labeled a, b, c, c.

a. Cut out the four slices, but before cutting, label each edge a, b, or c.

b. Try your hand at reconstructing the original square from the four slices.

c. Now rearrange the cake slices to form two squares. Hint: Note that each slice has two right angles. The other set of four right angles should be placed at the corners of a large square. You should now have a large outer square, the four cake slices, and an empty inner square as shown in Fig. 3.4b. Note that when the cake slices are rearranged to form two squares, Fig. 3.4b, which shows the solution, will be larger than Fig. 3.4a since the inner square is empty.

d. Next, express in terms of a, b, c the areas of the outer square A_L, the combined cake slices which equals the area of the original square cake A_C, and the small inner square A_S. Replace the expressions of part d into the equation:

$$A_L = A_S + A_C,$$

which expresses the fact that the outer area is the sum of the small empty area and the area of the original cake slices.

e. After some algebra, you should be able to derive the expression:

$$a^2 + b^2 = 2c^2 = (c\sqrt{2})^2$$

which looks like the Pythagorean theorem. Figure 3.4a shows the triangle within the original cake slices for which this expression pertains. Figure 3.4b places edge lengths on each edge to facillitate this construction.

3.3.4. Pythagorean theorem by way of an Islamic tiling

Here is another *Aha!* proof. Figure 3.5a shows a design, attributed to Annairizi of Arabia (circa 900) tiled by a number of white and gray squares [Sar]. Notice that the design is also tiled by large squares containing dissections of the white and gray squares. In Fig. 3.5b these three species of squares are shown to be squares on the sides of a right triangle with the large square being the square on the hypotenuse. By observing this design, it is evident that the area of the square on the hypotenuse is the sum of the areas of two smaller squares, and therefore the Pythagorean theorem is again proven. This design could have been made with any right triangle and the three squares on its sides. Figure 3.5b shows how the large square on the hypotenuse can be dissected to create the pieces that form the two smaller squares.

The Annairizi pattern can be created by stamping out the square on the hypotenuse as shown in Fig. 3.5a.

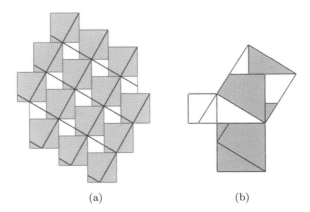

(a)   (b)

Fig. 3.5. Proof of Pythagorean theorem by Annairizi of Arabia

3.4. Examples of the Pythagorean Theorem in Islamic Art

Figure 3.6a shows the design of a ceramic mosaic found on the wall of a mosque in Isfahan circa 1700. It contains the name, in Kufic writing, of the designer-artisan who made it, in Kufic lettering. Figure 3.6b shows a schematic of this tile. You will notice that it is a square within a square. If you eliminate lines within the inner square as in Fig. 3.6c, you will see that it matches the *"Aha!"* proof in Sec. 3.3.2 above. If you eliminate the lines making up the outer square as in Fig. 3.6d you will notice the proof of the Pythagorean theorem from Sec. 3.3.1, attributed to Bhaskara. The creator of the Isfahan wall design chose dimensions of the sides of the triangles to be in the ratio of 2:1 although this is not necessary for the proof. With this 2:1 ratio, Fig. 3.6e shows that if the tiling is taken to be the square on the hypotenuse of a right triangle, the four triangles and the inner square can be juxtaposed to form the squares on the other two sides similar to what was done in the Annairizi proof.

We will now show how the Islamic designers used the Pythagorean proof based on four quadrilaterals (cake cuts) to create designs. However, to better understand this, we first digress to review some fundamentals of geometry.

3.5. Fundamentals of Geometry

a. Theorem 1: All inscribed angles drawn within a semicircle are right angles as illustrated in Fig. 3.7a.
b. Theorem 2: The perpendicular bisectors of the sides of a triangle meet at a point which is the center of the circumscribed circle as shown in Fig. 3.7b. If the triangle is a right triangle the meeting point lies on the hypotenuse as in Fig. 3.7c. If the triangle is a 45, 45, 90-right triangle the meeting point is at the center of the hypotenuse.

3.6. A Set of Designs by W.K. Chorbachi

We will now describe a series of designs discovered by Chorbachi pertaining to the cake cut proof of the Pythagorean theorem, described

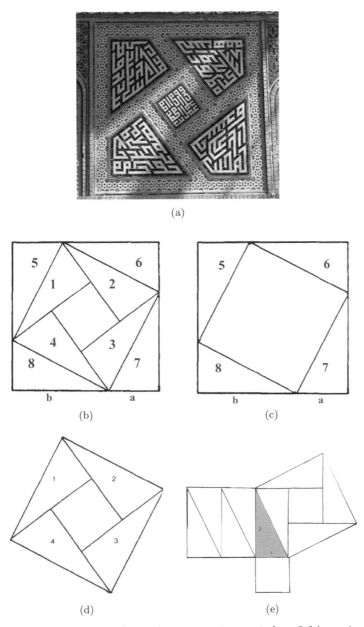

Fig. 3.6.   Pythagorean proofs based on a ceramic mosaic from Isfahan, circe 1700 by W.K. Chorbachi

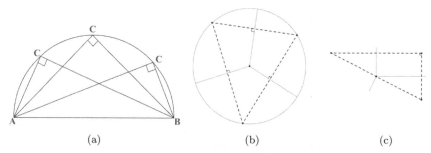

Fig. 3.7. (a) Right triangle in a semicircle; (b) Perpendicular bisectors of the sides of a triangle meet at a point; (c) For right triangles the perpendicular bisectors meet on the hypotenuse

in Sec. 3.3.3. She found this in an historical manuscript without title or author referred to simply as folio 192b at the Bibliotheque Nationale in Paris [Blo].

The design is based on an asymmetric quadrilateral, the shape of one cake cut from Fig. 3.4a with sides 2, 2, 1, $\sqrt{7}$. The quadrilateral is built up from a 45, 45, 90-triangle with compass and straightedge construction. Two adjacent quadrilaterals form an irregular space-filling pentagon with all sides equal to 2 units.

After creating the quadrilateral, we augment it to a similar quadralateral that matches Gardner's cake cut in Fig. 3.4a with sides a, b, c, c. We could carry out this procedure for any values of a, b, and c, but Chorbachi preferred a quadrilateral derived from a square with side $3 + \sqrt{7}$. We will also see that the proofs of the Pythagorean theorem on the wall ceramic from Isfahan illustrated in Fig. 3.6a is also represented in the square with augmented quadrilaterals.

We then reflect the augmented square on each of its four sides multiple times to form one of the seventeen spacefilling wallpaper patterns and study its symmetry. We find that the wallpaper pattern is identical to the spacefilling pentagons mentioned above.

Finally, we are able to convey some simple instructions to the artisans on how to turn this complex construction into design. We now describe this procedure step-by-step following the analysis of Chorbachi.

3.6.1. Derivation of the asymmetric quadrilateral

The derivation of an asymmetric quadrilateral follows from these steps:

a. Begin with a 45, 45, 90 right triangle with sides AB and AD equal to 2 units and hypotenuse BD equal to $\sqrt{8}$ (see Fig. 3.8a).
b. Draw a circumscribed circle about the midpoint of the hypotenuse (Fig. 3.8b) (see Theorem 2 in Sec. 3.5).
c. Mark off chord CD of unit length (Fig. 3.8b).
d. This defines two triangles ABD and BCD, both right triangles according to Theorem 1 in Sec. 3.5 (Fig. 3.8b).
e. The triangles are shaded in Fig. 3.8c.
f. This results in the asymmetric quadrilateral $ABCD$ (Fig. 3.8d) with sides $1, 2, 2, \sqrt{7}$.
g. Triangle BCD is singled out in Fig. 3.8e since this is the right triangle to which this proof of the Pythagorean Theorem construction refers.
h. Reflect quadrilateral $ABCD$ in a mirror line at BC to obtain a semi-regular pentagon with five equal sides of 2 units (Fig. 3.8f and 3.8g).

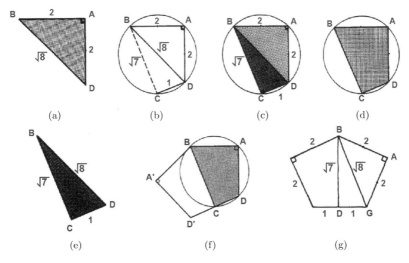

Fig. 3.8.   Derivation of the asymmetric quadrilateral by W.K. Chorbachi

The asymmetric quadrilateral $ABCD$ will be shown to be the key to the designs that Chorbachi discovered.

3.6.2. Scaling the asymmetric quadrilateral

In the next step we scale $ABCD$ to a larger similar rectangle as follows:

a. Begin with the quadrilateral $ABCD$ where $CD = 1$ unit, $AB = DA = 2$ units and $BC = \sqrt{7}$. (Fig. 3.9a).

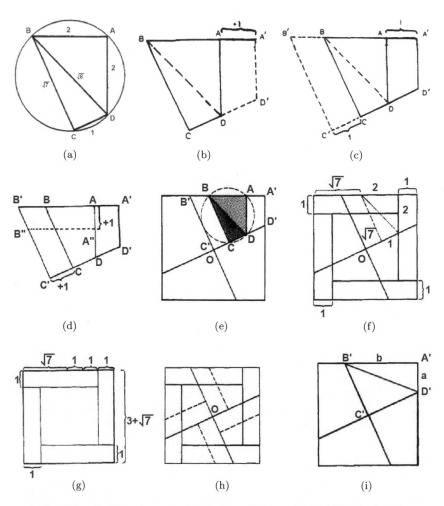

Fig. 3.9.   Scaling the asymmetrical quadrilateral by W.K. Chorbachi

b. Add the trapezoid $DD'A'A$ of width 1 unit to edge AD of quadrilateral $ABCD$ (Fig. 3.9b).

c. Add trapezoid $BB'C'C$ of width 1 unit to edge BC (Fig. 3.9c).

d. This results in quadrilateral $A'B'C'D'$ which is a scale model of the original (Fig. 3.9c). Note that $A'B'C'D'$, the scaled-up model of quadrilateral $ABCD$, is actually a mirror image and must be reflected and rotated (or cut out and turned upside down) to match the notation in Fig. 3.9c.

e. A final trapezoid $B'B''A''A$ of width 1 unit is added to edge AB' (Fig. 3.9d).

f. The two quadrilaterals, $ABCD$ and $A'B'C'D'$ are shown in relation to each other in Fig. 3.9e where the enlarged quadrilateral is shown to be one of Martin Gardner's cake cuts.

g. The other three cuts are gotten by rotating the first cake $D'C'B'A'$ cut through 90 deg., 180 deg., and 270 deg. about the center point O which is relabeled from C' (Fig. 3.9e).

h. The original quadrilateral $ABCD$ with some of the augmented trapezoids is shown in Fig. 3.9f. Notice, by observing the symmetry that the framing square has side $3 + \sqrt{7}$ (see Fig. 3.9f and 3.9g).

i. The framing square alone is shown in Fig. 3.9g.

j. The remaining trapezoids are added in Fig. 3.9h. Notice the ceramic wall mosaic of Fig. 3.6a with darkened symmetric quadrilaterals.

k. In Fig. 3.9i observe that $A'D' = a$, $A'B' = b$, and $B'C' = C'D' = c$ and $B'C' = d$ where a, b, c are the lengths of Gardner's cake cuts, and d is the diagonal of the right triangle specified by the cake cut problem (see Fig. 3.9i).

l. We can now determine the values of a, b, and c.

Quadrilateral $ABCD$ is similar to $A'B'C'D'$ since their corresponding angles are equal; in other words, they are scale models of each other. As a result, their sides are in proportion, i.e,

$$\frac{a}{1} = \frac{b}{\sqrt{7}} = \frac{c}{2} = \frac{d}{\sqrt{8}}.$$

Therefore, $b/a = \sqrt{7}$, $a + b = 3 + \sqrt{7}$, $c = 2a$, $d = a\sqrt{8}$.

Using algebra to solve for a, b, c and d results in the following values:

$$a = \frac{\sqrt{7} + 2}{3}, \quad b = \frac{2\sqrt{7} + 7}{3}, \quad c = 2a, \quad d = a\sqrt{8}.$$

Remark: When CD is taken to be 1 unit on the circumscribed circle of Fig. 3.9a, this results in the right triangle with $b/a = \sqrt{7}$. By choosing C to be a different point on the circumscribed semi-circle BCD, this would result in a different asymmetric quadrilateral.

3.6.3. Symmetry

In Fig. 3.9g the edges of the framing square are surrounded by mirrors.

The framing square with all of its auxiliary lines are reflected in the edges of the framing square. This is done repeatedly with the new squares to create a wallpaper pattern. Four such reproductions of the original square are shown in Fig. 3.10a. and this gives the starting point of a wallpaper symmetry design.

This wallpaper symmetry is referred to by Loeb [Loe 2] as 2 4 $\hat{4}$. This notation indicates that the wallpaper pattern is generated by 2-fold (180 deg.) and 4-fold (90 deg.) rotations labeled in Fig. 3.10a by 2 and 4. The 2-fold rotations lie at the intersection of the vertical and horizontal mirrors while the 4-fold rotations lie on lines of glide reflection (see Sec. 14.1).

In Fig. 3.10b the set of symmetric quadrilaterals are darkened. These are similar to the symmetric quadrilaterals in the classic figure at Isfahan shown in Fig. 3.6a. In fact, one of these symmetric quadrilaterals subjected to four quarter turns leaves an empty square and results in a diagram similar to the classic diagram found at Isfahan (see Fig. 3.6a).

Observe in Fig. 3.10c that the wallpaper pattern in Fig. 3.10a can be represented as the semi-regular pentagons constructed in Fig. 3.8f and Fig. 3.8g as shown in Fig. 3.10c, Fig. 3.10d also shows that the symmetric quadrilateral occurs at three different scales in the wallpaper symmetry half-turn about the center of the edge.

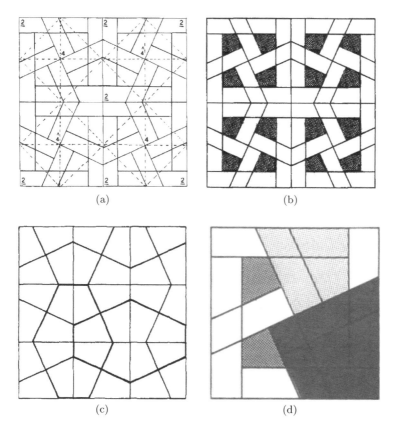

Fig. 3.10. Symmetry of the asymmetric quadrilateral by W.K. Chorbachi

3.6.4. A set of designs based on the asymmetric quadrilateral

We now apply this construction to creating some complex tilings as an Islamic artisan would.

a. Figure 3.11a shows that the asymmetric quadrilateral tiles the plane by itself as a wallpaper symmetry $22'2''2'''$, using Loeb's notation [1971] in which each edge is given a half-turn about the center of the edge.

b. Notice in Fig. 3.11b that the perpendicular bisectors of each of the sides of the quadrilateral meet at a common point, the center of the circumscribing circle that contains the four vertices of the

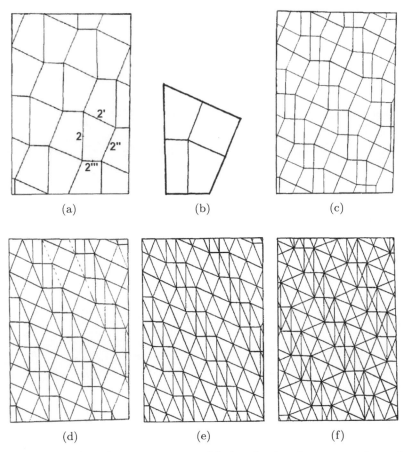

Fig. 3.11. An algorithm to create a set of designs bardon the asymmetric quadrilateral by W.K. Chorbachi

quadrilateral. These perpendicular bisectors divide the quadrilateral into a square, a rectangle, and two congruent asymmetric quadrilaterals similar to the original.

c. In Fig. 3.11c the tiling with asymmetric quadrilaterals are shown with their perpendicular bisectors.
d. In Fig. 3.11d left leaning diagonals are added to the asymmetric quadrilaterals.
e. In Fig. 3.11e left leaning diagonals are inserted in the rectangles and squares.

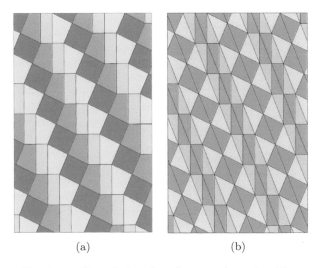

(a)   (b)

Fig. 3.12.   Two designs based on previous algorithm

f. Finally, in Fig. 3.11f, right leaning diagonals are added to the rectangles and squares.

g. This results in a "game board" in which countless designs can be created by color-coding the various regions in this tiling as shown for the two designs in Fig. 3.12a and 3.12b by Kevin Miranda.

3.7. Conclusion

We have seen how the creation of designs can be reduced to the carrying out of a series of algorithms that may have been executed by artisans leaving the geometrical details to the mathematicians. The fact that a significant mathematical idea, that of the Pythagorean theorem, was involved in the creation of the designs of this paper, contributes to the interest of the designs. Clearly, the Islamic mathematicians were knowledgeable of advanced geometrical principles in order to carry out such designs.

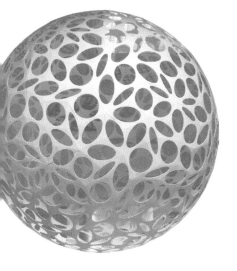

Chapter 4

Tiling a Rectangle by Congruent and Non-congruent Squares

4.1. Introduction

We first tile a rectangle by congruent squares and then develop a more complex method to tile a rectangle by non-congruent squares. When a square is tiled by non-congruent squares, this is also known as "squaring the square" or a "perfect square dissection." The first squared rectangle was discovered in 1909 by Z. Moron in which Moron found a 33 by 32 rectangle which uses nine square tiles of different sized edge lengths. However, for years, mathematicians claimed that a perfect square dissection was impossible. However, in 1936 four students at Trinity College — R.L. Books, C.A.B. Smith, A.H. Stone, and W.T. Tuttle succeeded in creating the first squared square using 69 tiles. Later, this was reduced by A.W.J. Duivestijn to 21 tiles [Pic].

4.2. Tiling a Rectangle with Congruent Squares

A rectangle with sides in proportion 3:2 is shown in Fig. 4.1. Clearly, the fewest number of congruent squares that are needed to tile it are six squares of side 1 unit. On the other hand the fewest number of congruent squares needed to tile a 15:9 rectangle is fifteen squares measuring 3 units on a side (see Fig. 4.2).

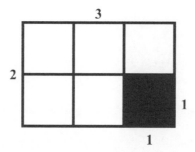

Fig. 4.1.   A 3 × 2 rectangle

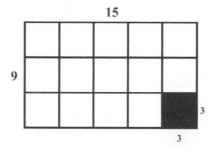

Fig. 4.2.   A 15 × 9 rectangle

We now ask the following question: For a rectangle of proportion $b : a$, where a and b are integers, what are the fewest number of congruent squares needed to tile the rectangle and what is the side of the square?

The answer is that the side of the smallest triangle is the greatest common divisor of a and b, i.e., gcd(a, b). By definition, the greatest common divisor of integers a and b, gcd(a, b), is the largest integer that divides evenly into both a and b. The number of squares N required to tile the rectangle is then,

$$N = \frac{a}{\gcd(a, b)} \times \frac{b}{\gcd(a, b)}. \tag{4.1}$$

Clearly, for 3 and 2, gcd(3, 2) = 1 whereas gcd(15, 9) = 3.

Remark: In Section 1.5 we found that gcd$(n, k) = 1$, i.e., n and k should be relatively prime was the condition that guarantees that a star polygon can be drawn in a single stroke without taking the pencil off of the paper. Two integers are said to be *relatively prime* if their gcd is 1.

Another quantity that is important to the theory of numbers is the least common multiple of a and b, i.e., $\text{lcm}(a, b)$ where $\text{lcm}(a, b)$ is the smallest integer which can be divided evenly by both a and b. It can be shown that

$$a \times b = \text{lcm}(a, b) \times \gcd(a, b). \tag{4.2}$$

From Eqs. 4.1 and 4.2 it follows that the minimum number of tiles N can be elegantly expressed by the following formula:

$$N = \frac{\text{lcm}(a, b)}{\gcd(a, b)}. \tag{4.3}$$

For example, since $\text{lcm}(3, 2) = 6$ and $\gcd(3, 2) = 1$, it follows from Eq. 4.3 that the number of tiles needed to tile the 3×2 rectangle is $N = 6$. Since $\text{lcm}(15, 9) = 45$ and $\gcd(15, 9) = 3$, it follows that $N = 15$.

What if you have a pair of large numbers such as the 60×27 rectangle shown in Fig. 4.3. How do you find the gcd? You can find the gcd by extracting squares. First you can extract two 27×27 squares with 6 units left over. Next extract four 6×6 squares, and finally two 3×3 squares can be extracted. The side length of the smallest square is the $\gcd\{60, 27\} = 3$. This procedure is equivalent

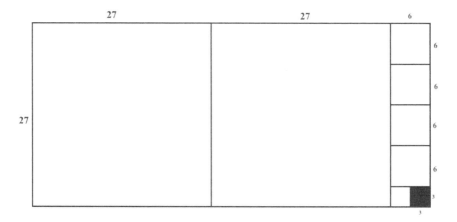

Fig. 4.3. Using rectangles to compute the greatest common denominator (gcd)

to what is called the *Euclidean algorithm* in the subject of Discrete Mathematics. If a and b are relatively prime, the side of the smallest square is 1 unit.

Problem: Use this method of extraction to find the gcd of the following rectangles: (a) 240:72 (b) 55:34. Find the least number of congruent squares needed to tile the 32×33 rectangle.

4.3. Tiling a Rectangle with Non-congruent Squares

It was quite easy but uninteresting to tile a rectangle by congruent squares when the sides are integers. It is more interesting to now consider the tiling of a rectangle by a finite set of squares, no two of which have the same edge length, i.e., non-congruent squares.

Since the set of squares is finite, there is a smallest square. The first problem we must confront in our tiling is where to place this smallest square. Can you see why it would be impossible to place the smallest square in a corner or along one of the edges of the rectangle as shown in Fig. 4.4a and 4.4b? The only other possibility is to place it within the interior of the rectangle as shown in Fig. 4.4c.

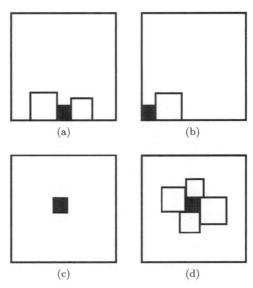

Fig. 4.4. (a–d) Non-congruent squares: Finding where to place the smallest square

Next we must decide how to surround the smallest square with other squares. Fig. 4.4d shows how this square must be surrounded. (why?)

Now that we have decided what to do with the smallest square we can show how the rest of the tiling can be determined. Consider a rectangle cut into smaller rectangles in such a way that there is a chance of distorting all the small rectangles into squares. In doing this we must be sure that at least one rectangle (the one we distort into the smallest square) is surrounded by four rectangles as in Fig. 4.4d. One candidate with nine rectangles is shown in Fig 4.5.

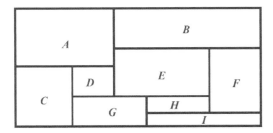

Fig. 4.5. Tiling a rectangle by rectangles

Either D or H could become the smallest square in this tiling. Let us choose H to be the smallest square and assign it an edge length of y units while assigning x units to square E. Fig. 4.6 shows a step by step assignment of side lengths to all of the other squares of the tiling in terms of x and y.

We now have an array of rectangles whose interiors have been labeled in terms of x and y so as to indicate the lengths of the sides of the squares when the rectangles are distorted to become squares. Since the left and right sides and top and bottom of the rectangle must be equal, this leads to two equations:

$$(x + 11y) + (x + 7y) = (2x + y) + (x + 2y) + (x + y) \quad (4.4a)$$

$$(x + 11y) + (2x + y) = (x + 7y) + (x + 3y) + (x + 2y). \quad (4.4b)$$

Solving Eq. 4.4a for x and y yields $x = 7y$, while solving Eq. 4.4b yields an identity. Therefore we may choose any value for y to determine the dimensions of all squares in this arrangement. If $y = 1$, then

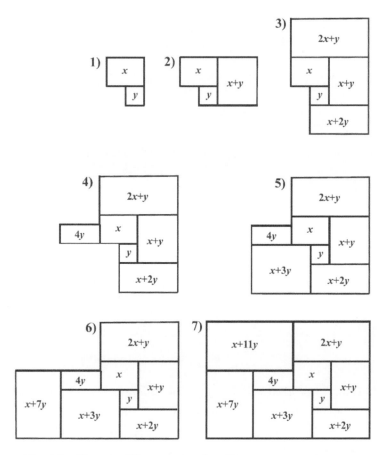

Fig. 4.6.   Steps to tiling a rectangle with non-congruent squares

$x = 7$ and the sides of the other squares can be determined to be: $A = 18$, $B = 15$, $C = 14$, $D = 4$, $E = 7$, $F = 8$, $G = 10$, $H = 1$, and $I = 9$. We have therefore determined the solution that is now drawn to scale in Fig. 4.7. The square tiling of the 32×33 rectangle of Moron's, has been created.

This is a beautiful arrangement that was not achieved by guesswork. It is the result of mathematical reasoning.

Notice that tilings can be turned into graphs. We do this by assigning letters to different heights within the tiling. For example in Fig. 4.7 there are six different heights labeled from a to f. Each of these heights becomes a vertex in a graph. The edge of the graph then

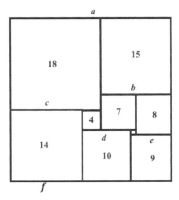

Fig. 4.7. The result of tiling a rectangle with non-congruent squares

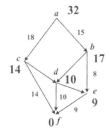

Fig. 4.8. An electrical analogy for the tiling of Fig. 4.7

becomes the side length of the square between two heights (vertices). In other words, there are as many vertices as heights in the tiling, and as many edges as the squares in the tiling. The graph corresponding to Fig. 4.7 is shown in Fig. 4.8. There is an electrical analogy here. If the side lengths are thought of as voltages, then voltages can be assigned to each vertex (using larger integers) where edges in the graph correspond to changes of voltage from one vertex to another. In this way, vertex 'a' has a voltage equal to the length of the left and right sides of the rectangle, or 32. The voltage at level b is then 17 and level c is 14, etc. while the voltage at the bottom of the square is 0, or ground.

Construction: Figures 4.9a and 4.9b show two other arrangements of 9 distorted squares. Distort these configurations so that the outer rectangle is tiled by non-congruent squares. First, assign letters and

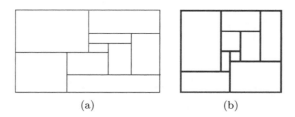

Fig. 4.9.   (a) and (b) Tile these rectangles with non-congruent squares

then decide which rectangle will be transformed into the smallest square, then follow the procedure outlined above. After determining the size of the squares, carry out the tiling with the care of an artist or designer. Color your tilings using the fewest number of colors possible so that no two squares sharing the same edge are the same color. Draw the graphs corresponding to these tilings. Can you rearrange the squares in your tiling so as to tile the outer rectangle in a different way?

Problem: Figures 4.10a and 4.10b are tilings of a 112×75 rectangle by 13 non-congruent squares for two tilings with the same set of squares. Draw their graphs.

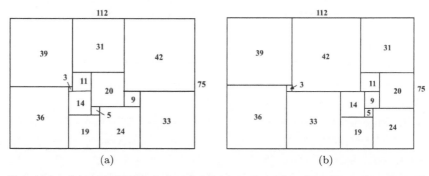

Fig. 4.10.   (a) and (b) Find the electrical analogy for this tiling of a 112×75 rectangle with non-congruent squares

Remark: There are no solutions to the tiling of a rectangle by non-congruent squares with fewer than 9 squares. There are two solutions with exactly 9 squares, six with 10 squares (four of which can be obtained by annexing a square to one side of a 9-square solution);

and 22 with 11 squares (12 of which are a direct result of 10-square solutions).

Try your hand at discovering a tiling not represented here. Be forewarned that many diagrams will result in impossible results; both variables will be equal to 0, or one variable or variable expression will be negative. Try working backwards from the graph to the tiling.

Problems: Two additional arrangements of distortions of tilings with 10 squares are given in Figs. 4.11a and 4.11b. Find the edge lengths of the squares and draw their graphs.

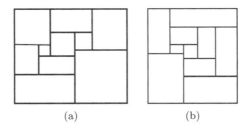

(a)   (b)

Fig. 4.11. (a) and (b) Tile these two rectangles with non-congruent squares

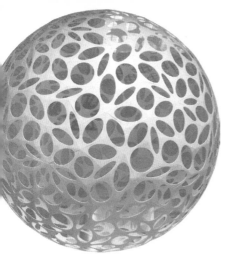

Chapter 5

Simple Tilings with Lattice Symmetry

5.1. Introduction

We present a simple method of obtaining two classes of lattice symmetry patterns in both two and three dimensions. In other words the patterns are invariant under translation in two or three independent directions. In the plane these patterns constitute two of the 17 classes of wallpaper patterns. This approach was developed by William J. Gilbert and appeared in Structural Topology No. 8 in 1983 [Gil]. We make use of his examples and add some of our own.

5.2. Lattices

A 2-dimensional *lattice* is a set of points that are unchanged or *invariant* under translations in two non-parallel directions as illustrated in Fig. 5.1. This lattice is characterized by three positive integers, k, h, and l and vectors $\vec{v_1} = (k, 0)$ and vector $\vec{v_2} = (h, l)$ where $0 \leq h < k$. Taking a linear combination of these two vectors, i.e., $a\vec{v_1} + b\vec{v_2} = (ak + bh, bl)$ where a, b are integers, both positive and negative, yielding the lattice shown in Fig. 5.1. A shaded square is placed at the initiating point of each lattice point.

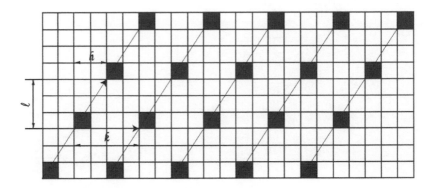

Fig. 5.1. A 2-dimensional lattice

5.3. A Lattice Tiling with Only Translational Symmetry

We illustrate this for the case where $k = 4$, $l = 3$, and $h = 2$, subdivided into 3×4 rectangles known as *fundamental domains* where in this case the *fundamental domain* is a rectangle as shown in Fig. 5.2.

a. Number the squares in this fundamental rectangle by the integers 1 to 12. Your design can only be made within these 12 integers. Box 1 is shaded and reveals the lattice.
b. Your design is made by coloring these 12 boxes. However, the boxes do not have to all be in a single rectangle but can be spread

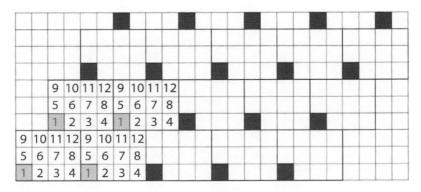

Fig. 5.2. A fundamental domain for a 2-D lattice with $k = 4$, $l = 3$, and $h = 2$

across several rectangles. However, there are no repeats and all 12 numbers must be used.

c. We refer to the pattern formed by these 12 numbers as the *fundamental design* pattern shown shaded in Fig. 5.3a and outlined in 5.3b.

d. Finally, color several fundamental designs, with the designs varying in color. This completes the pattern which is invariant only under translations in the two directions given by the lattice vectors.

1	2	3	4	1	2	3	4
11	12	9	10	11	12	9	10
7	8	5	6	7	8	5	6
3	4	1	2	3	4	1	2
9	10	11	12	9	10	11	12
5	6	7	8	5	6	7	8
1	2	3	4	1	2	3	4

(a)   (b)

Fig. 5.3. (a) A fundamental pattern associated with the fundamental domain; (b) Outline of fundamental patterns

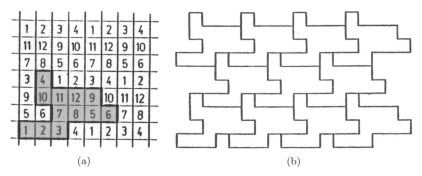

Fig. 5.4. The fundamental pattern in Fig. 5.3a and two patterns with minor modifications

Once you have a design, it is easy to make minor modifications as illustrated by the shaded patterns in Fig. 5.4. Three designs characterized by $k = 10$, $l = 1$ and $h = 3$ are shown in Fig. 5.5a and 5.5b.

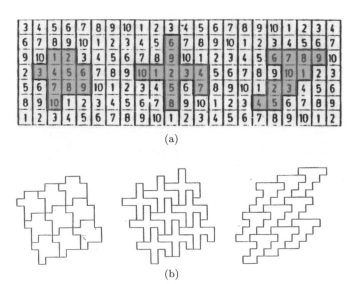

(a)

(b)

Fig. 5.5. (a) A three fundamental patterns with $k = 10$, $I = 1$ and $h = 3$; (b) Outline of the 3 fundamental patterns from Fig. 5.5a

5.4. Lattice Symmetry Patterns with 4-fold Rotations

Consider a sequence of lattice squares formed by the vectors $(k, 0)$ and $(0, k)$ shown in Fig. 5.6a. Next notice that the lattice points form a square tiling over the lattice where the *fundamental square* is shaded. Do you notice that points A and B in the fundamental square are points of 4-fold rotation? In other words, if you rotate the tiling by 90, 180, 270 degrees about these points the tiling does not change, or as we say, it is invariant. The tiling is also invariant under 2-fold rotations about point C, i.e., the tiling does not change under a half-turn about point C. Next, divide the fundamental square (shaded) into small squares numbered 1 through 9 as shown in Fig. 5.6b. If you color these numbers by various colors you get what is known as the fundamental pattern. Then rotate this fundamental pattern by 90, 180, and 270 deg. to get the lattice square shown in Fig. 5.6b shown in bold print. Finally, this lattice square can be translated in the directions of the two vectors characterizing the lattice to number all points in the plane. Notice that the pattern that emerges has centers A, B, C of 4-fold and 2-fold symmetry as shown above.

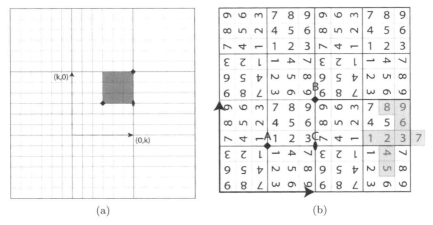

| (a) | (b) |

Fig. 5.6. (a) A lattice for a tiling with 4-fold rotational symmetry; (b) A fundamental domain for a lattice with 4-fold rotational symmetry

Remark: It will always occur that the sum of the inverses of centers of n-fold rotations in the fundamental domain add up to 1, e.g., in the fundamental square the centers are at points A, B, and C therefore for our fundamental square,

$$\frac{1}{4} + \frac{1}{4} + \frac{1}{2} = 1.$$

5.5. Another Lattice Symmetry Pattern with 4-fold Rotations

Half squares will also tile the plane as shown in Fig. 5.7a. Half-squares pervade Japanese design where it manifests as the shape of the so-called *tatami mat*. This pattern of squares, subdivided into half-squares, is invariant under translations by vectors $(k, 0)$ and $(0, k)$, shown in Fig. 5.7b, where k is a positive even integer. In other words, this pattern has a lattice structure. It can then be seen that the half-squares are $k/2 \times k$ rectangles. The fundamental rectangle is shaded in Fig. 5.7a. Within the fundamental rectangle points A and B are centers of 4-fold rotation while point C is a center of 2-fold rotation.

In Fig. 5.7b we let $k = 8$ in which case the plane is tiled by 8×8 squares subdivided into 2×4 and 4×2 half-squares. The fundamental

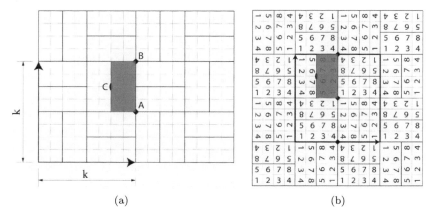

(a)   (b)

Fig. 5.7. (a) Another lattice 4-fold rotational symmetry; (b) Another fundamental domain for a lattice with 4-fold rotational symmetry

rectangle has dimensions 4×2 and the boxes in this rectangle are numbered 1 to 8. If these numbers are assigned varying colors, this results in the fundamental pattern. Using the 2-fold and 4-fold centers of rotation, all of the boxes inside of an 8×8 square, with darkened edges, are then assigned integers from 1 to 8. These integers can be spread to the entire plane by using the translation vectors of the lattice.

5.6. Three Dimensional Lattice Tilings

To extend the concept of lattice symmetry to three dimensions, we need a third vector. Fig. 5.8a illustrates a lattice system with three vectors characterized by five integers: k, l, m, h, i, j where,

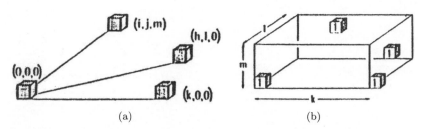

(a)   (b)

Fig. 5.8. (a) Three vectors defining a 3-dimensional lattice; (b) A fundamental domain for a 3-D lattice

$\vec{v_2} = (k,0,0)$, $\vec{v_2} = (h,l,0)$, and $\vec{v_3} = (i,j,m)$. Instead of a fundamental rectangle we now have a *fundamental parallelepiped* as shown in Fig. 5.8b. The parallelepiped is then subdivided into $k \times l \times m$ cubes numbered from 1 to *klm*. The pattern is then made by choosing all *klm* cubes with no repeats. Again, all cubes do not have to lie in one parallelepiped.

As an example, we let $k = 6$, $l = 2$, $m = 1$, $h = 4$, $i = 1$, and $j = 1$. The numbers 1 through 12 are taken on two levels. The resulting structure is shown in Fig. 5.9a and more explicitly in Figs. 5.9b and 5.9c.

A second structure is defined by $k = 5$, $l = 3$, $m = 1$, $h = 1$, $i = 3$, $j = 1$. The numbers 1 through 15 are distributed over three parallelepipeds as shown in detail in Fig. 5.10a and Fig. 5.10b and more explicitly in Fig. 5.10c.

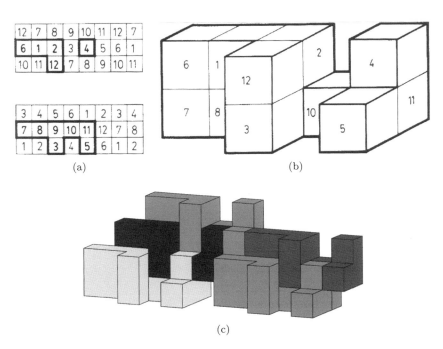

Fig. 5.9. (a, b, c) Steps to a three dimensioal design with lattice symmetry

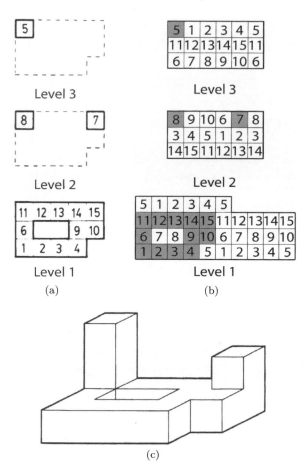

Fig. 5.10. (a, b, c) Steps to another design with 3-dimensional lattice symmetry

5.7. Examples of Lattice Tilings

In Fig. 5.11 three lattice tilings are represented along with their fundamental patterns. Additional tilings can be found on the website.

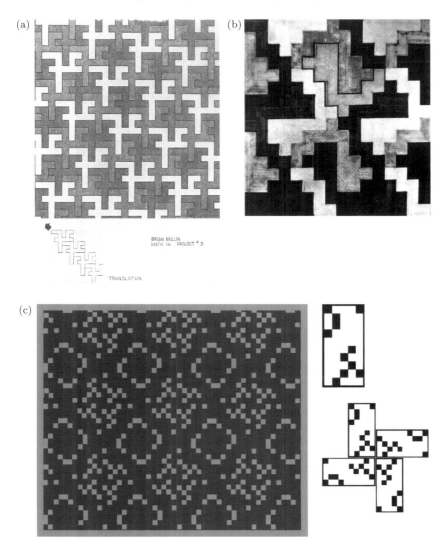

Fig. 5.11. (a) Example of a patten with lattice symmetry and only translations showing the fundamental pattern; (b) Example of a pattern with lattice symmetry showing 4-fold rotation; (c) Another pattern with 4-fold lattice symmetry showing the fundamental pattern.

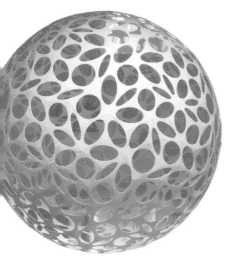

Chapter 6

The Brunes Star

The Danish engineer, Tons Brunes, has studied the properties of the eight pointed star shown in Fig. 6.1. He claims to have seen it on tapestries found in Pompeii and Hertzegovina. He felt that it was one of the templates at the basis of ancient temple construction and other sacred structures. I saw this so-called *Brunes star* made up of string in Barcelona at the home of the Spanish architect, Antonio, Gaudi and also on the ceiling of the gift shop at the great Gaudi cathedral in Barcelona, Sagrada Familia. Its structure is both elementary and surprising.

To construct the Brunes star, divide a square into two half-squares by a vertical and a horizontal line and place the two diagonals into each of these four half-squares. What emerges is the Brunes star with its internal structure, as shown in Fig. 6.2. The star has been subdivided entirely into (3, 4, 5)-triangles or fragments of such triangles at four different scales. The star can also be constructed by taking four string loops subdivided into 12 equal parts. Each loop can be bent into a (3, 4, 5)-triangle anchored with pins. When these loops are placed properly into a square, the star emerges (see the emboldened triangle in Fig. 6.2).

This star has many interesting properties described in my book Beyond Measure [Kap2] It gives an approximate squaring of the circle in both length and area. Finally, when its two diagonals are placed

Fig. 6.1.   A Brunes star

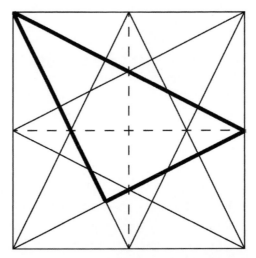

Fig. 6.2.   One of four 3, 4, 5-right triangles that underline the structure of a Brunes star

in the original square, Fig. 6.3 shows that a line segment placed at different heights is divided by the star into 1, 2, 3, 4, 5, 6, 7, 8 equal parts. The line segment is divided into seven equal parts at the level of the sacred cut (see Fig. 2.3). as shown in Fig. 6.4. The square is

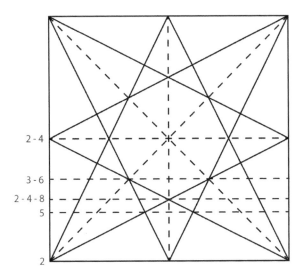

Fig. 6.3. Examples of how the Brunes star equipartitions a horizontal line

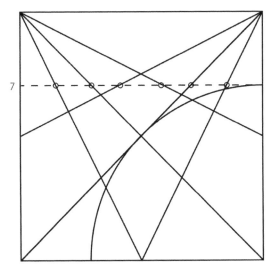

Fig. 6.4. The sacred cut determines the level at which a horizontal line is equipartitioned into 7 parts

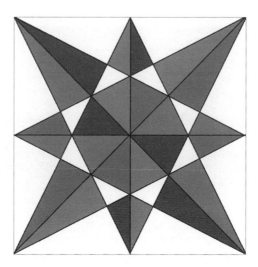

Fig. 6.5.   One example of a Brunes star design by a Macarena Maldonado

now divided into four subsquares each able to recreate a Brunes star at a smaller scale. This can then be continued ad infinitum.

Construction: Create a Brunes star with its internal structure and color the regions to bring out its power. One design from my studio on Math of Design by Macarena Maldonado is shown in Fig. 6.5. Other designs can be found on the website.

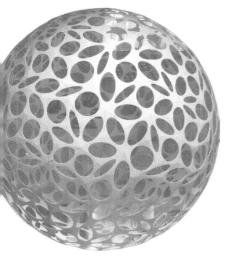

Chapter 7

Do You Like Paleolithic Op-art?
by Slavik Jablan

7.1. Introduction

In this chapter we consider the history, principally from Eastern Europe [Jab3, Jab5], of certain modular elements: versatiles, op-tiles, Kufic tiles, and key patterns, which occur as ornamental archetypes from Paleolithic times until the present. In the next chapter we will go into detail as to how these elements can be implemented in the classroom, and in Chap. 9 we will discuss applications. The use of modularity in design is emphasized where a simple design technique is used over and over showing how complexity follows from simplicity. A particular focus will be made to a set of modules involving black and white stripes drawn on a square that we refer to as *versatiles*.

7.2. Paleolithic Ornaments

The scholar of symmetry, knot theory, and the history of design, Slavik Jablan, found that the oldest examples of ornamentation in Paleothic art were from Mezin (Ukraine) dated to 23 000 B.C. Note that 23 000 years is a time period ten times longer than the complete written history of mankind. At first glance, the ornament on the right side of Fig. 7.1a appears to not be significant; it is an ordinary set of parallel lines. On the right side of Fig. 7.1b this pattern is transformed into a set of parallel zig-zag lines — an ornament with a symmetry group of type 244 (see Sec. 3.6.4). Using the notation of Arthur Loeb, generated by an axis of reflection perpendicular to another axis of glide reflection (Fig. 7.1b). Let's see how the creative process for the design of this ornament may have developed. Imagine a modern engineer who begins a construction project. At first, he makes a rough sketch, and then he begins to work more seriously to solve the problem.

The next series of ornaments from Mezin is more advanced. The previously mentioned sets of parallel lines are arranged in friezes and meander patterns that can still be considered as sketches (Fig. 7.1c, d).

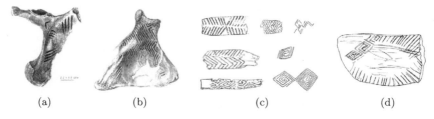

(a)   (b)   (c)   (d)

Fig. 7.1.   Basic patterns from Mezin

In Figure 7.2a we see the final result, the masterpiece of Paleolithic art — the Birds of Mezin decorated by meander ornamentation. The man of prehistory has applied the symmetry constructions that he learned, and he has preserved them for posterity. On the mammoth bone, modelled in the form of a bird, he engraved the meander pattern which represents the oldest example of a rectilinear spiral in the form of a meander ornamentation.

(a)   (b)   (c)

Fig. 7.2.   (a) Bird of Mezin; (b) Mezin bracelet; (c) developed bracelet

The next artifact is an engraved bracelet from the same excavation site (Fig. 7.2b). As you can see, this bracelet is carved from a mammoth bone only two millimeters thick. If you try to engrave such an ornament on the tiny bone layer, you will not succeed since it will break. Therefore, the man of prehistory made this bracelet with the expenditure of much time and effort in the following way: first he cut a portion of the mammoth bone, then on its surface engraved the ornamentation, removed the internal part of the bone, and he obtained the engraved bracelet. If we look at this bracelet in developed form (Fig. 7.2c), we notice that there is a continuous transition from one ornament to another via a third ornament: on the left corner; you can see the meander ornamentation, then the set of parallel zig-zag lines used as a symbol of water, and again the continuous transition to another meander ornamentation. In order to make a continuous transition from one ornament to the next, it is necessary to have a relatively high level of the mathematical knowledge and precision, which is unexpected for Paleolithic times [Jab3].

How is the continuous transition from one ornament to another made? The ornaments on Fig. 7.3 look very different one from another. Among them are black-white and colored ornaments, and at first glance, it appears that there is no unifying principle. Their common property is that they all consist of a single element (module). Notice the small black-white square in the middle. It consists of a set of parallel diagonal black and white stripes. If this square is used as the basic motif, then all of these ornaments can be constructed from it. We call this method of construction the principle of modularity. Our goal is to construct all ornaments or structures by using the smallest number of basic elements (modules) and to obtain, by their recombination, as many different ornaments (structures) as possible.

This module, a square or rectangle with a set of parallel diagonal black and white stripes, is the so-called versatile as we will see in the next chapter. It is the basis of Mezin meander patterns (Fig. 7.2). In the next chapter we will describe these striped elements from the point of view of creating such patterns in the classroom or by

Fig. 7.3. Modular key-patterns

designers. In Chap. 9 we will show the application of versatiles to meanders and meander knots.

Figure 7.4 shows the ornaments from Scheila Cladovei culture (Romania, 10 000 B.C.) based on similar modules — rectangular Op-tiles.

Fig. 7.4. Ornaments from Scheila Cladovei culture (Romania, 10 000 B.C.)

7.3. Neolithic Ornaments

Figure 7.5 shows a series of ornaments from Titsa culture (Hungary) and Vincha (Serbia), dating to 3 000–4 000 B.C. They are painted on ceramic and can be found in similar Neolithic settlements. How

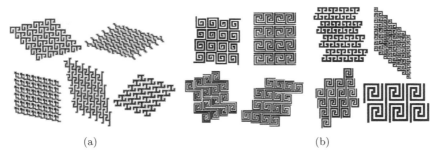

(a)   (b)

Fig. 7.5.   Neolithic ornaments from (a) Titsa culture (Hungary) and (b) Vincha (Serbia)

did the Neolithic people come to the idea of constructing such ornaments? We will try to show that all ornaments were derived from the simplest of human technologies: basketry, weaving, matting, plaiting, or textiles. Then the best of ornaments (in an aesthetic sense) were copied to the stronger media of bone, stone, and ceramics. Many of these ornaments are obtained from interlaced patterns (fabrics). If we take two bands of different colors and make the simplest possible interlacing pattern: "over-under", "over-under", ... we obtain the anti-symmetric ("black-white") checkerboard pattern. By replacing the simple code: "over-under", "over-under", ... by a more sophisticated code (a repetitive algorithm), we obtain more complex and visually more interesting interlacing patterns [Rad1]. Notice that ornaments from Vincha (Fig. 7.5b) are all based on meanders, continuing the tradition of Paleolithic ornaments from Mezin and Scheila Cladovei (Fig. 7.4).

By observing the numerous examples of Neolithic "black-white" ornaments, with the black part ("figure") congruent to the white part "ground", we conclude that all of them originated from basketry, matting, plaiting, weaving, or textiles and then were copied to the stronger media of stone, bone and ceramic (see Fig. 7.6).

In this sense, it is very convincing to observe that the left image in Fig. 7.7a shows a well dressed Neolithic man wearing a dress with stripes which is similar to the module used for the construction of the ornaments from Mezin [Jab3]. In Fig. 7.7b there are Neolithic figurines, the first from Vincha, the second from the Titsa region

Fig. 7.6.   Neolithic ornaments on ceramics: Tisza (Hungary), Cucuteni (Romania), Vincha (Serbia), Dimini (Grece), Titsa and Miskolc (Hungary), Serra d'Alto (Italy), Rakhmani (Greece)

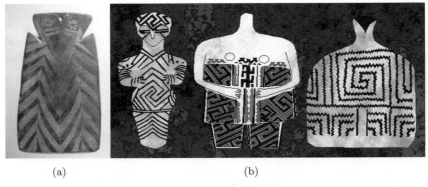

(a)   (b)

Fig. 7.7.   Neolithic plate from Portugal; (b) Neolithic figurines from Vincha (Serbia), Titsa region (Hungary), and Vadastra (Romania)

(Hungary), and the third from Vadastra (Rumania), where on parts of their clothes appear very similar types of meander ornamentation transferred from one culture to the other. This is also a testament to the notion that textile ornamentation was used as a model for similar ornamentation on ceramic.

The best textile patterns were copied to ceramic vessels which requires great skill since the surface of the ceramic vessels are curved. We can find similar examples all over the world (e.g., in Neolithic Lapita ceramics from Fiji (Fig. 7.8a), or Anasazi ceramics (Fig. 7.8b)).

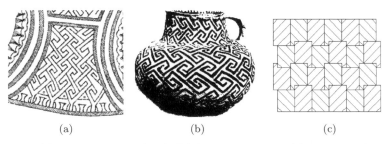

(a) (b) (c)

Fig. 7.8. (a) Lapita ceramics (Fiji); (b) Anasazi ceramics; (c) Example of a tiling with overlap

7.4. Kufic Tiles

If in a black and in a white square we construct one diagonal region of opposite color, we obtain two modular elements called Kufic tiles. From these modular elements, named by S. Jablan "Kufic tiles" [Jab4], you can create square Kufic letters and Kufic scripts: the writing of letters, names, or texts, where all black and white lines are of the same width. The Kufic tile is the simplest versatile: a white square with one black diagonal stripe, or its negative. The following example shows the name of Allah (Alli) written in ornamental Kufic script (Barda, Azerbaijan, XIV century). By using Kufic tiles, we can (re)construct ornaments from the Alhambra (e.g., the famous "Maple leaf" pattern in Fig. 7.9b) or make a Kufic script logo for

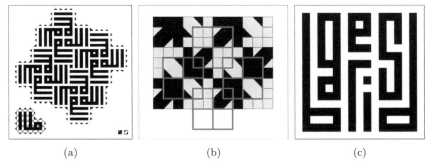

(a) (b) (c)

Fig. 7.9. (a) Ornamental Kufic script (Barda, Azerbaijan, XIV century); (b) reconstruction of the "Maple leaf" ornament from Alhambra; (c) "Bridges" in Kufic script

the "Bridges Conference" (Fig. 7.9c, design by Lj. Radovic and S. Jablan).

Kufic tiles can be also used for simple constructions of Islamic ornaments and Moorish patterns from the Alhambra (Granada, Spain) [Sar]. In this case, overlapping of tiles is permitted as you can see from the "Maple leaf" tiling constructed using the complete set of Kufic modules (Fig. 7.9). In the next chapter we show how Kufic writing can be implemented. Professor Donald Knuth, a master of computer art and the author of the program *TeX*, designed new TeX-fonts from our Kufic tiles.

7.5. Labyrinths

The word "labyrinth" is derived from the Latin word *labris*, meaning a two-sided axe, a motif related to the Minos palace in Knossos. The walls of the palace were decorated by these ornaments while the interior of the palace features bronze double axes. This is the origin of the name "labyrinth" and the famous legend about Theseus, Ariadne, and the Minotaur [Fen1]. The Cretan labyrinth is shown on the silver coin from Knossos (400 B.C.) (Fig. 7.10a).

How does one construct a unicursal maze? Figure 7.10 shows the most elegant way: draw a black meander (Fig. 7.10b), remove several rectangles or squares, rotate each of them around its center by the 90 deg. angle, and place it back to obtain a labyrinth (Fig. 7.10c). Even very complex mazes can be constructed in this way (Fig. 7.11).

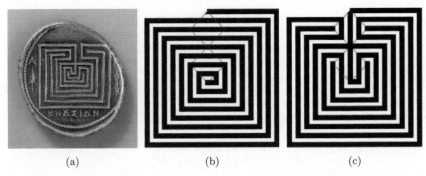

(a)   (b)   (c)

Fig. 7.10.   (a) Silver coin from Knossos with the image of a labyrinth; (b) a spiral meander which can be composed of Versatiles; (c) its transformation to maze

Fig. 7.11. Construction of a more complicated maze

In Chap. 9, we will present more about the history of labyrinths and how they can be created.

7.6. Op-art

Op-tiles, which first appeared in Mezin (Ukraine, 23 000 B.C.), are related to Op-art (optical art) from the end of twentieth century where they were abundantly used. Representative examples are "Square of Three" by R. Neal (1964) (Fig. 7.12), "Hyena stomp"

Fig. 7.12. "Square of Three" by R. Neal (1964)

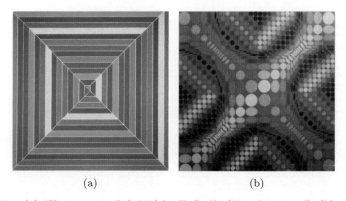

(a)   (b)

Fig. 7.13.   (a) "Hyena stomp" (1962) by F. Stella (Tate Liverpool); (b) variation on the theme of Op-tiles by V. Vasarely

(1962) by F. Stella (Fig. 7.13a), with colored Op-tiles producing a meander, and a variation on the theme of Op-tiles by V. Vasarely (Fig. 7.13b) [Bar1, Jab6]. In the next chapter we present patterns that can be used to create Op art.

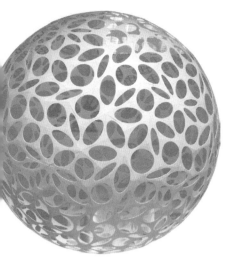

Chapter 8

Truchet, Versatiles, Op-art and Kufic Tiles by Slavik Jablan and Ljiljana Radovic

8.1. Introduction

The last chapter illustrated some of the history, in large part in Eastern Europe, of design going as far back as 23,000 B.C. in Mezin in the Ukraine. This chapter will focus on how to implement the elements of this historical design, which include Versatiles, optiles, Truchet tiles and Kufic tiles.

8.2. Versatiles

Great versatility in the creation of designs can be achieved from the simplest of origins. As the last chapter showed, Slavik Jablan traced the creation of modular designs through history. In his work, Jablan feels that clothing design may have provided early impetus through the creation of a simple system of diagonal stripes placed on a square as shown in Fig. 8.1. Models of these so-called *versatiles* are shown in Fig. 8.2. If you focus on the left tile and the right tile you will notice that in the first tile, to the left, there is a simple system of black and white stripes. In the second tile to the right, black becomes white and white becomes black. The remaining three tiles

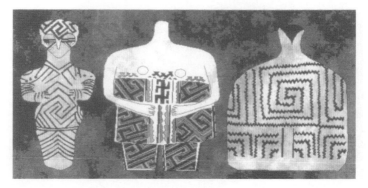

Fig. 8.1.   Neolithic artefacts from Vincha (Serbia), Tisza (Hungary) and Vadastra (Romania)

Fig. 8.2.   Versatiles and Op-tiles (author S. Jablan)

represent simple tiles that lead to op-art designs and are referred to as *optiles* [Jab3].

In Fig. 8.3, several designs from my studio on Mathematics of Design are shown. In Chap. 7, which Jablan entitled, "Do You Like Paleolithic Op-Art?," he gave many examples of versatile designs that he has traced as far back as 23,000 B.C. [Jab5].

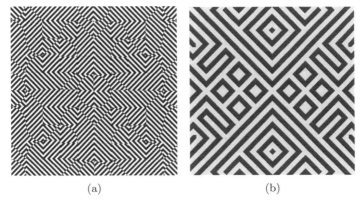

Fig. 8.3. (a) Versatile design by Jonathan Martin; (b) Op-tile design by Jonathan Martin

Fig. 8.4. A spiral built from overlapping square versatiles

These versatiles can be generalized by using rectangles instead of squares. Also the squares need not be with their edges one beneath the other, but they can be moved relative to each other. This is done in Fig. 8.4 to form spirals. Also the tiles can overlap. Finally, you can take the spirals and by giving several tiles 90 deg. rotations, you can create designs that reproduce labyrinths such as the Cretan Labyrinth shown in Fig. 8.5 [Jab7]. We will return to this system in Chap. 9 on meanders.

Fig. 8.5.   Two square versatiles are rotated 90 deg. to create the Cretan Labyrinth

8.3.   Truchet Tiles

A second modular tile is the *Truchet* tile which is simply black and white half-squares separated along the diagonal shown in Fig. 8.6. Using only four tiles, you can create many different designs. Increasing to 25 tiles, you can create countless designs such as the ones in

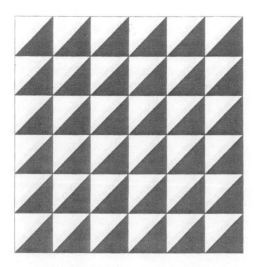

Fig. 8.6.   A set of Truchet tiles

Fig. 8.7.   Designs with Truchet tiles

Fig. 8.7. And this is the beauty of the modular approach, diversity follows from simplicity [Jab3, Rad2].

8.4. Kufic tiles

A third set of tiles are referred to as *Kufic* tiles since they lead to a form of calligraphy known as Kufic writing [Jab3]. Again, they are created by bisecting the edges of a square and connecting the bisection points to form a hexagon as shown in Fig. 8.8a. Again, these tiles come in pairs where white is changed to black and black to white as shown in Fig. 8.8b. Figure 8.9 shows the beginning of how the system of black and white Kufic tiles can join together to create Kufic writing using only the white-black and black-white Kufic tiles and the black isosceles triangles which appear in the corners of the Kufic tiles. The creation of two letters are shown in Fig. 8.10. This can be best carried out by using the computer graphics program, Inkscape. Finally, a complex result is shown in Fig. 8.11.

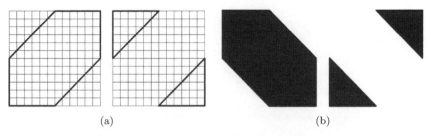

(a)   (b)

Fig. 8.8.   (a) A pair of Kufic tiles created on a square grid; (b) a pair of black and white Kufic tiles

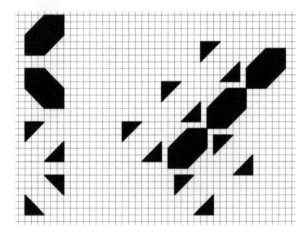

Fig. 8.9.   Steps to Kufic writing

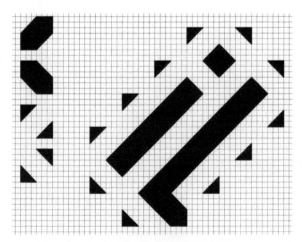

Fig. 8.10.   Two Kufic letters are formed

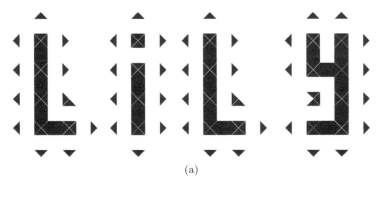

(a)

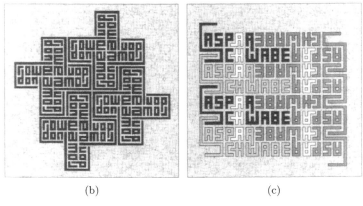

(b)   (c)

Fig. 8.11.   (a) The letters spell the name LiLy by Kevin Miranda; (b) and (c) are two additional examples of Kufic writing by Slavik Jablan.

Construction 1: Construct a design using versatiles or optiles. You may enlarge, copy and cutout the templates for versatile and optiles, in Fig 8.12 to create your designs.

Construction 2: Create a spiral and then try turning the spiral into a labyrinth by rotating several elements of the spiral.

Construction 3: Create some interesting designs using the Truchet tiles.

Construction 4: Try to create your name with Kufic writing. Examples are shown in Figs. 8.11a, 8.11b and 8.11c.

Fig. 8.12.   Templates for versatiles and optiles.

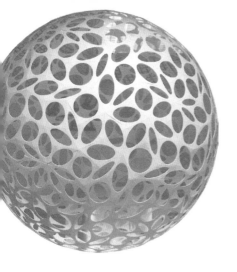

Chapter 9

Meanders, Knots, Labyrinths, and Mazes by Jay Kappraff, Slavik Jablan, Ljiljana Radovic, Kristof Fenyvesi

9.1. Introduction

The meander motif got its name from the river Meander, a river with many twists mentioned by Homer in the Iliad and by Albert Einstein in a classical paper on meanders [Ein]. The motif is also known as the Greek key or Greek fret shown in Fig. 9.1 with other Greek meander patterns. The meander symbol was often used in Ancient Greece, symbolizing infinity or the eternal flow of things. Many temples and objects were decorated with this motif. It is also possible to make a connection of meanders with labyrinths since some labyrinths can be drawn using the Greek key. We will refer to any set of twisting and turning lines shaped into a repeated motif as a *meander pattern* where the turning often occurs at right angles [LaCroix, 2003]. For applications of meanders, the reader is referred to [Arn], [DiF]. This chapter is in large measure, a reworking of an earlier paper by Kappraff, Jablan and Radovic [Kap9] and two other earlier papers by Jablan and Radovic [Jab1] and Jablan, Radovic and Fenyvassen [Fen].

Perhaps the most fundamental meander pattern is the meander spiral which can be found in very early art history. The prototile

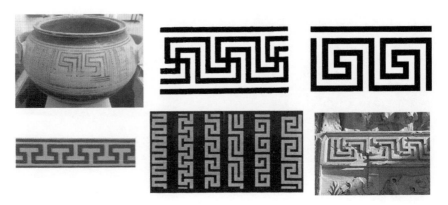

Fig. 9.1.   Greek meanders

based on a set of diagonal stripes drawn on a square and a second square in which black and white are reversed, called versatiles, described in the last chapter, are used abundantly in ornamental art going back to Paleolithic times. From these two squares an infinite set of "key patterns" can be derived. The commonly found patterns were independently discovered by various cultures (Paleolithic, Neolithic, Chinese, Celtic), and were independently discovered by these cultures. The oldest examples of key-patterns are ornaments from Mezin (Ukraine) about 23,000 B.C. [Jab3]. The appearance of meander spirals in prehistoric ornamental art can be traced to archeological findings from Moldavia, Romania, Hungary, Serbia and Greece, and all of them can be derived as modular structures. In this chapter we will study the application of meanders to frieze patterns, labyrinths, mazes and knots.

9.2. Meander Friezes

To create a *frieze pattern* begin with a basic pattern and translate the pattern along a line in both directions. There are seven classes of frieze patterns employing reflections in a mirror along the line, mirrors perpendicular to the line, and half turns at points along the line. One of each of the seven frieze patterns is shown in Fig 9.2. A subclass of meander friezes can be formed from the initiating pattern within a $p \times q$ rectangular grid of points as shown in Fig. 9.3. A continuous set of line segments is placed in the grid touching each

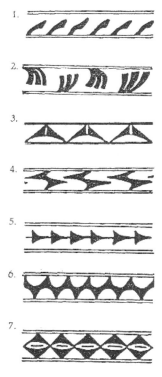

Fig. 9.2.   Examples of the seven frieze patterns

point with no self-intersections. If the grid points are considered to be vertices and the line segments are edges of a graph, then such a path through the graph is referred to as a *Hamiltonian path*. In this way, the pattern has numerous twists and turns inducing a meander configuration, resulting in what we refer to as a *meander frieze*. The pattern has one edge that enters the grid and another leaving the grid at the same level in order to connect to the next translated pattern.

The number of frieze patterns corresponding to each square tends to be quite large. Even a 5 × 5 grid gives (up to symmetries) 19 different cases as shown in Fig. 9.4 and the 7 × 7 grid gives more than 2800 possibilities. The variety of meander friezes can be further enriched by inserting some additional internal elements (intersections), for example, a rosette with a swastika motif as shown in Fig. 9.5. It is clear that Ancient Greeks and other cultures created friezes using only a very small portion of the possibilities, which are

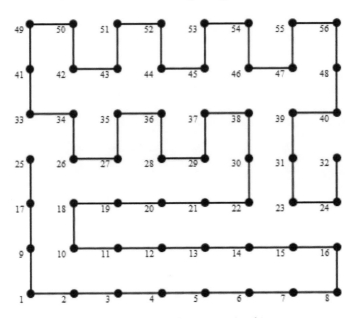

Fig. 9.3.   Pattern for a meander frieze

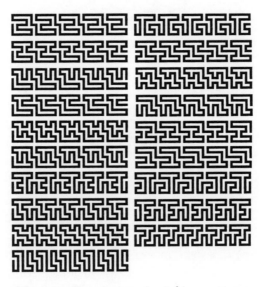

Fig. 9.4.   The nineteen 5 × 5 frieze patterns

Fig. 9.5. A rosette with a swastika motif

Fig. 9.6. A 7 × 7 frieze pattern

restricted only to grids of small dimensions. Hence, meander friezes originating from grids of dimension 7 × 7 such as the pattern in Fig. 9.6 were probably not used at all.

9.3. Meanders Represented by the Intersection of Two Lines

The creation of meander patterns is based on the notion of an open meander [Bevan (2013)].

Definition 1: An *open meander* is a configuration consisting of an oriented simple curve, and a line in the plane referred to as the axis. The simple curve crosses the axis a finite number of times and intersects only transversally [LaCroix, 2003].

In this way, open meanders can be represented by systems formed by the intersections of two curves in the plane. Two meanders are equivalent if one can be deformed to the other by redrawing it without changing the number and sequencing of the intersections. In this case the two meanders are said to be *homeomorphic*. They occur in the physics of polymers, algebraic geometry, mathematical theory of mazes, and planar algebras, in particular, the Temperly-Lieb algebra. One such open meander is shown in Fig. 9.7a. As the main source of the theory of meanders, we used the paper [LaC]. For applications of the theory of meanders, the reader is referred to [LaC], [Arn], [DiF].

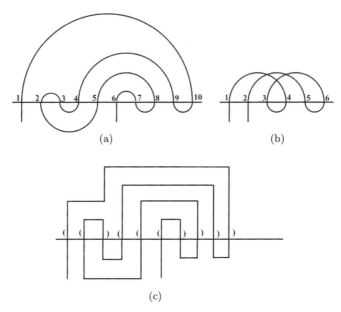

Fig. 9.7. (a) Open meander given by meander permutation $(1, 10, 9, 4, 3, 2, 5, 8, 7, 6)$; (b) non-realizable sequence $(1, 4, 3, 6, 5, 2)$; (c) piecewise-linear arch configuration given by Dyck word $\{(()((()))); 1(())1()()\}$

The order of a meander is the number of crossings between the meander curve and the meander axis. For example, in Fig. 9.7a there are ten crossings so the order is 10. Since a line and a simple curve are homeomorphic, their roles can be reversed. However, in the enumeration of meanders we will always distinguish the meander curve from the meander line, referred to as the axis. Usually, meanders are classified according to their order. One of the main problems in the mathematical theory of meanders is their enumeration.

An open meander curve and meander axis have two loose ends each. Depending on the number of crossings, the loose ends of the meander curve belong to different half-planes defined by the axis for open meanders with an odd order, and to the same half-plane when the meanders have an even order. For example, in Fig. 9.7a, the loose ends are in the same half-plane since it has an even order. In this case, we are able to make a closure of the meander: to join each of the loose ends. We will find that an odd number of crossings results in a *knot* whereas an even number of crossing results in a *link*. *Knots* consist of a single strand whereas *links* are characterized by the interlocking of multiple strands. We will discuss knots in Sec. 9.4.

We will use *arch configurations* to represent meanders.

Definition 2: An *arch configuration* is a planar configuration consisting of pairwise non-intersecting semicircular arches lying on the same side of an oriented line, arranged such that the feet of the arches are a piecewise linear set equally spaced along the line as shown in Fig. 9.7a.

Arch configurations play an essential role in the enumeration of meanders. A meandric system is obtained from the superposition of an ordered pair of arch configurations of the same order with the first configuration as the upper and the second as the lower configuration.

The modern study of this problem was inspired by [Arnol'd, 1988]. If the intersections along the axis are enumerated by $1, 2, 3, \ldots, n$ every open meander can be described by a *meander permutation* of order n, the sequence of n numbers describing the path of the meander curve. For example, the open meander in Fig. 9.7a is coded by the *meander permutation* $(1, 10, 9, 4, 3, 2, 5, 8, 7, 6)$. Enumeration of open meanders is based on the derivation of meander permutations which play an important role in the mathematical theory of labyrinths [Phi]. In every meander permutation, odd and even numbers alternate, i.e, parity alternates in the upper and lower configurations. However, this condition does not completely characterize meander permutations. For example, the permutation: $(1, 4, 3, 6, 5, 2)$ exhibits two crossing arches, $(1, 4)$ and $(3, 6)$ as shown in Fig. 9.7b. Therefore, the most important property of meander permutations is that all arches must be nested in order not to produce crossing lines. Among different techniques to achieve this, the fastest algorithms for deriving meanders are based on encoding each configuration as words in the Dyck language [Jensen, 1999; Bobier, 2013] and the Mathematica program "Open meanders" by David Bevan (https://demonstrations.wolfram.com/OpenMeanders/) [Bevan, 2013]. The upper and lower arches are represented by nested parentheses with the loose ends represented by 1's. As a result, the upper and lower arches in Fig. 9.7a are coded by: $\{(()((()))), 1(())1()()\}$. The nested curves can also be "squared off" as shown in Fig. 9.7c.

9.4. Meander Knots

First, we say a few words about knots [Ada]. A knot can be thought of as a knotted loop of string having no thickness. It is a closed curve in a space that does not intersect itself. We can deform this curve without permitting it to pass through itself, i.e., no cutting. Although these deformations appear quite different as shown in Fig. 9.8, they are considered to be the same knot. If a deformation of the curve results in a simple loop, it is referred to as an unknot. To create the *shadow* of a knot, draw a scribble of lines with the restriction that at any point of intersection only two lines of the scribble intersect as shown in Fig. 9.9a. Notice that at each point of intersection of the

Fig. 9.8.   A knot and its deformations

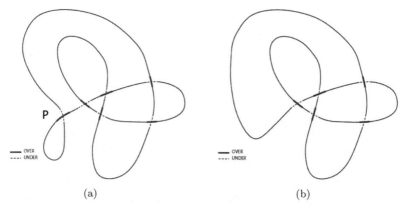

(a)   (b)

Fig. 9.9.   (a) The shadow of a knot resulting in a knot with one extraneous crossing at P: (b) the knot redrawn with the crossing at P removed

scribble four edges intersect. By introducing the over/under relation in crossings of the shadow, we get a knot diagram. An *alternating knot* can be constructed from its shadow by drawing a path through the scribble, entering a point of intersection and taking the middle segment of the three exit choices and then proceeding along the path in an over-under-over-under — ... pattern as shown in Fig. 9.9b. Notice that some crossings, such as the crossing at point P can be eliminated by simple twists or movements without cutting. These moves are referred to as *Reidemeister* moves of which there are three such unknotting rules [Adams, 1994]. After all such movements are made, the resulting knot can be reduced to its minimum number of crossings as shown in Fig. 9.9b. A minimal projection of a knot is one that minimizes the number of crossings. This is called the *crossing number*, defined to be the least number of crossings that occur in any projection of the knot. It is uniquely defined for any knot. Minimal Meander diagrams have a minimal number of crossings.

As we described in Sec. 9.3, when making a closure of meander diagrams we have two possibilities. We choose the one producing a meander knot shadow without crossing lines (loops). After that, by introducing under-crossings and over-crossings along the meander knot shadow axis, we can turn it into a knot diagram. When the crossings alternate: under-over-under-over ... the knot is said to be *alternating*. Given a knot, it can be transformed without cutting to eliminate certain crossings. However, unless the knot is a loop or unknot there will always remain crossings, specified by the crossing number. Each knot can be classified by its *crossing number*.

Definition 3: An alternating knot that has a minimal diagram in the form of a minimal meander diagram is called a *meander knot*.

Another problem is the derivation of meander knots first introduced by S. Jablan. Several meander knots are represented here by their *Gauss codes* and Conway symbols [Con]; [Rol], [Jab2]. All computations were obtained by Jablan using the program "Lin-Knot" [Jab3]. Gauss codes of *alternating meander knot diagrams* can be obtained if to the sequence $1, 2, 3, \ldots, n$ we add a meander permutation of order n where n is an odd number and in the obtained sequence alternate the signs of successive numbers, e.g., from meander permutation $(1, 8, 5, 6, 7, 4, 3, 2, 9)$ we obtain Gauss code $\{-1, 2, -3, 4, -5, 6, -7, 8, -9, 1, -8, 5, -6, 7, -4, 3, -2, 9\}$ which corresponds to the alternating knot with nine crossings referred to by 9_7 in the literature, and also given by the Conway symbol 3 4 2. The same knot can be obtained from the meander permutations: $(1, 8, 7, 6, 5, 2, 3, 4, 9)$ and $(1, 8, 7, 4, 5, 6, 3, 2, 9)$ with Gauss codes $\{-1, 2, -3, 4, -5, 6, -7, 8, -9, 1, -8, 7, -6, 5, -2, 3, -4, 9\}$, and $\{-1, 2, -3, 4, -5, 6, -7, 8, -9, 1, -8, 7, -4, 5, -6, 3, -2, 9\}$.

These three representatives of the knot 3 4 2 are shown in Fig. 9.10. It should also be pointed out that if an alternating knot has an alternating minimal meander diagram, all of its minimal diagrams need not be meander diagrams.

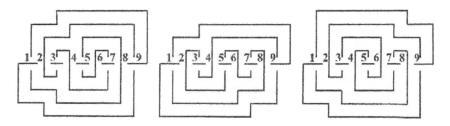

Fig. 9.10.   Non-isomorphic minimal meander diagrams of the knot $9_7 = 342$

The natural question which arises is to find all alternating meander knots with n crossings where n is an odd number [Rad3]. Table 9.1 specifies the number of Open Meanders (OM) with n crossings and the number of their corresponding Alternating Meander Knots (AMK) with at most $n = 9$ crossings are illustrated in Fig. 9.11.

Which knots can be represented by non-minimal meander diagrams? For example, the figure-eight knot, in Conway notation 22 with four crossings, cannot be represented by a meander diagram but can be represented by the non-minimal meander diagram given by the Gauss code $\{-1, 2, -3, -4, 5, 3, -2, 1, 4, -5\}$, a knot with five crossings. Knot 22 and five additional non-minimal diagrams are shown in Fig. 9.12. You will also note that the knot is not alternating. For every knot which is not a meander knot (does not have a minimal

Table 9.1. The number of open meanders and alternating meander knots with n crossings

n	OM	AMK
1	1	1
3	2	1
5	8	2
7	42	5
9	262	15
11	1828	52
13	13820	233
15	110954	1272

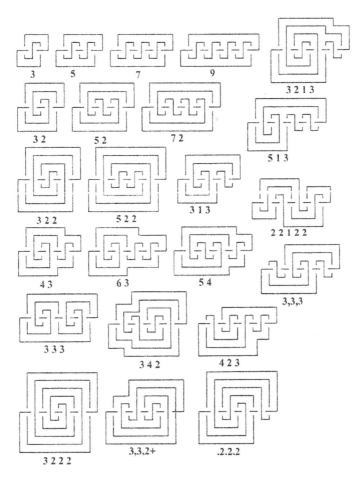

Fig. 9.11. Alternating meander knots (AMK) with at most $n = 9$ crossings

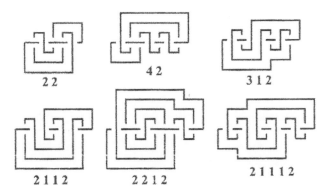

Fig. 9.12.   Non-minimal meander diagrams of knots:$4_1, 6_1, 6_2, 6_3, 7_6, 7_7$.

meander diagram) but which can be represented by some meander diagram (which is reduced, but has more crossings than the minimal diagram of that knot, i.e., more crossing than the crossings number of that knot), we can define its meander number, the minimum number of crossings of its meander diagrams where the minimum is taken over knots with the same shadow in which some crossings are changed from overcrossing to undercrossing and vice versa. The next step is to make all possible crossing changes in alternating minimal diagrams, i.e., in Gauss codes of alternating meander knots and see which knots will be obtained.

9.5. Two Component Meander Links

Open meanders with an even number of crossings offer the interesting possibility of joining pairs of loose ends of the meander axis and the meander curve. As a result, we obtain the shadow of a 2-component link with one component in the form of a circle and the other component meandering around it. A natural question is which alternating links can be obtained from these shadows and, in general, which 2-component links have meander diagrams. It is clear that components do not self-intersect so the set of 2-component meander links coincides with the set of alternating 2-component links with non-self-intersecting components, and all of their minimum diagrams preserve this property.

As for knots, we pose for 2-component links the natural question as to which 2-component links have meander diagrams. From the definition of meander links it is clear that the answer will be links in which both components will be knots and in which components are not self-crossing, i.e., it will be the shadow of a circle. In the case of alternating minimal meander diagrams, all such diagrams of 2-component links will have this property. However, in the case of non-minimal meander diagrams, some links with an odd number of crossings are represented by meander diagrams. Moreover, their minimal diagrams have components with self-intersections, but in their non-minimal meander diagrams none of the components have self-intersections. Meander links up to $n = 10$ crossings are shown in Fig. 9.13.

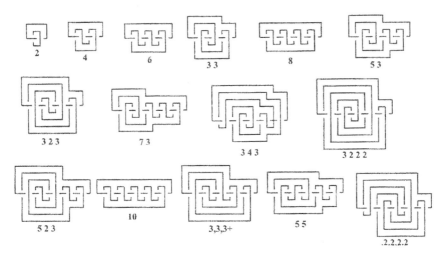

Fig. 9.13.   Meander links up to $n = 10$

9.6. Labyrinths

According to the Greek myths, the skillful craftsman Daedalus created the Labyrinth. The purpose of this special architectural structure was to imprison the Minotaur, the son of Pasiphae, the wife of the Cretan King Minos. The myth of the Cretan Labyrinth has been a subject of speculation and an archaeological, historical, and anthropological research for a long time — just as the visual representations of labyrinthine structures concern not only art historians, but also mathematicians [Fen2].

Karl Kerenyi (1897–1973), the internationally renowned scholar of religion — colleague of Carl Jung, and friend and advisor of Thomas Mann, returned time after time to the mythological research of labyrinths and interpreted them both as cultural symbols and specific geometrical structures. Right from the beginning of his labyrinth studies, Kerenyi introduced the labyrinth from three closely interrelated main aspects: (1) as a mythical construction; (2) as a spiral path that was followed by dancers of a specific ritual; and (3) as a structure that was represented by a spiral line. In his 1941 essay series [Ker1], he summarized the most important concepts of previous studies and made several original observations and comparisons which are still widely quoted and referred to in Labyrinth Studies. With the comparative mythological and morphological analysis of the Babylonian, Indonesian, Australian, Norman, Roman, Scandinavian, Finnish, English, German and medieval and Greek labyrinth tradition, he has proven the global presence of labyrinthine structures and revealed the artistic and architectural impulse behind the creation of them to rituals and cultic dances where participants followed a spiral line and made meandering gestures and dance movements. In 1963, Kerenyi devoted a lengthy essay to Greek folk dance [Ker3] and pointed out how the movements of the ancient labyrinth dances were transformed into the main components of the Syrtos, a dance that is still performed in Greece today. And in his last book written in 1969 [Ker2] where he explored the Cretan roots of the cult of Dionysus, he discussed in depth the labyrinthine and meander-like patterns of Knossos in dance, art, and architecture.

"When a dancer follows a spiral whose angular equivalent is precisely the meander, he returns to his starting point", wrote Kerenyi, quoting Socrates from Plato's dialogue *The Euthydemus*. Socrates speaks there of the labyrinth and describes it as a figure whose most easily recognizable feature is an endlessly repeated meander or spiral line: "Then it seemed like falling into a labyrinth; we thought we were at the finish, but our way bent round and we found ourselves, as it were, back at the beginning, and just as far from that which we were seeking at first [Ker3]. There resulted a classical picture of this procession which originally led by way of concentric circles and surprising turns to the decisive turn in the center where one was obliged to rotate on one's own axis in order to continue the circuit [Ker1]. The labyrinths "surprising turns" and the "decisive turn in their center" is responsible for their symbolic meaning as well. Kerenyi sees the labyrinth as a depiction of Hades, the underworld, and interprets the structures as narrative symbols which express the existential connection between life and death, between the oblivion of the dead and the return of the eternal living.

From a morphological perspective, Kerenyi presupposes the transformation of the spiral to the meander pattern because straight lines were easier to draw and so the rounded form was early changed into the angular form. For Kerenyi, the meander is the figure of a labyrinth in linear form. In the third to second centuries BC, as he explains, we find the figure and the word unmistakably related: in the Middle Ages labyrinths were also called meanders [Ker3]. We find a detailed connection between meanders and labyrinths in Matthews book, *Mazes and Labyrinths* [Mat].

Although both Matthews and Kerenyi made the connection between labyrinths and meanders clear, the ornamental evolution of angular labyrinths were not discussed by any of them in a way that could explain the geometrical development process underlying them. Our approach seeks to remedy this. Before proceeding I would like to make clear the difference between labyrinths and mazes since these words are often used interchangebly. Both labyirinths and mazes can be described by graphs. However, in the case of labyrinths, there is a single path leading from the entrance to the center, whereas for mazes there are, at various points, bifurcations in the path with some choices of continuance leading to dead ends and others leading on to the center. So in a sense labyrinths can be thought of as being subsets of mazes in which there is a *unicursal path* through the graph.

9.7. Labyrinth Studies and Visual Arts

We have found that the oldest examples of geometrical ornamentation in Paleolithic art were from Mezin (Ukraine) dated to 23,000 B.C. (see Fig. 9.14) [Jab3].

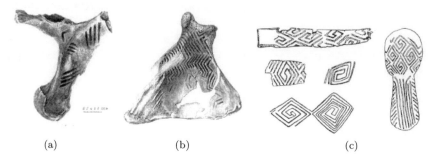

(a)   (b)   (c)

Fig. 9.14.   Ornaments from Mezin (Ukraine) circa 23,000 B.C.

Among the set of ornaments found at Mezin is the first known meander frieze under the well-known name "Greek key." Take a set of parallel lines, cut a square or rectangular piece with the set of diagonal parallel lines incident to the first ones, rotate by 90 deg., and if necessary translate it in order to fit with the initial set (See Fig. 9.15). More aesthetically pleasing results will be obtained by using the initial set of black and white strips of equal thickness.

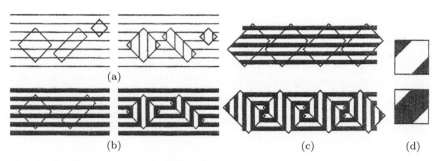

(a)

(b)   (c)   (d)

Fig. 9.15.   Meander patterns formed from parallel strips with rotated squares and rectangles by the "cut and paste" method

9.8. From Meanders to Labyrinths

The word "labyrinth" is derived from the Latin word labris, making a two-sided axe, the motif related to the Minos palace in Knossos. The walls of the palace were decorated by these ornaments while the interior featured actual bronze double axes. This is the origin of the name "labyrinth" and the famous legend about Theseus, Ariadne, and the Minotaur [Fen1]. The Cretan labyrinth is shown on the silver coin from Knossos (400 B.C.) as shown in Fig. 7.10a and 9.16.

To create the Cretan labyrinth, first consider a Simple Alternating Transit labyrinth or SAT labyrinths [Arn]; [Phi]. An SAT labyrinth is laid out on a certain number of concentric or parallel levels. The labyrinth is *simple* if the path makes essentially a complete loop at each level, in particular, it travels on each level exactly once. It is *alternating* if the labyrinth -path changes direction whenever it changes level and *transit* if the path progresses without bifurcation from the outside of the maze to the center [Jab3, Fen2].

Fig. 9.16. The Cretan labyrinth

Most SAT labyrinths occur in a spiral meander form with the path leading from the outside to the center. Each such labyrinth can be sliced down its axis and unrolled into an open meander form. Now the path enters at the top of the form and exits at the bottom: the top level (center) of the labyrinth becomes the space below the open meander form. This process is illustrated in Fig. 9.17 for the Cretan labyrinth. The topology of an SAT labyrinth is entirely determined by its level sequence, i.e., its open meander permutation as described in Sec. 9.3; for example, the meander permutation {3,

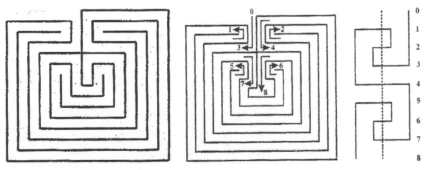

Fig. 9.17. Meander permutation and the "unrolling" procedure to create the Cretan labyrinth

2, 1, 4, 7, 6, 5} is illustrated in Fig. 9.17. Hence, the enumeration of open meanders and their corresponding SAT labyrinths is based on the derivation of meander permutations. For the derivation of open meanders one can use the Mathematica program "open meanders" by David Bevan [2013] which we modified in order to compute open meander permutations.

How does one construct a unicursal path without knowledge of computer programs and topological transformations? The simplest natural labyrinth is a spiral meander: a piecewise-linear equidistant spiral. It is defined by a simple algorithm: start from a central point and after every step turn by 90 deg, and continue with the next step, where the sequence of step distances is 1, 1, 2, 2, 3, 3, 4, 4 .... Tracing this sequence we have a labyrinth path: a simple curve connecting the beginning point (the entrance) with the end point (Minotaur room). Fig. 9.18 shows an elegant way to construct a Cretan maze. Draw a black spiral meander (Fig. 9.18a), cut out several rectangles or squares, rotate each of them around its center by 90 deg., and place it back to obtain a labyrinth (Fig. 9.18b).

Even very complex labyrinths can be constructed in this way (Fig. 9.19a and b). It is interesting to notice that the Knossos "dancing pattern", using the shape of a double axe, can be reconstructed in a similar way (Fig. 9.19c and d). So, a simple pattern (Fig. 9.20), an optile [Jab3] can be considered as the logo of a Paleolithic designer from which Mezin ornaments can be created. These tiles were also discovered by Ben Nicholson [Nic] who referred to them as Versatiles.

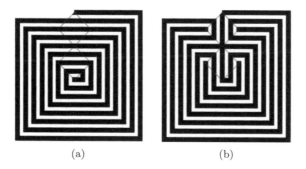

(a)   (b)

Fig. 9.18.   (a, b) "Cut and paste" construction of a complex labyrinth

(a)   (b)

(c)   (d)

Fig. 9.19.   (a, b) A spiral meander; (c, d) The Knossos "dancing pattern"

Fig. 9.20.   Prototiles

9.9. A Labyrinth Workshop

1. From linoleum squares of dimension 40×40 cm and self-adhesive tape of two colors (e.g. black and silver) make the basic tiles (Fig. 9.21).
2. Make a meander spiral (Fig. 9.22). By rotating only two tiles by 90 deg. you obtain the Cretan Labyrinth as was done by S. Jablan at the Bridges. Conference in Pecs 2010 shown in Fig. 9.22. Notice that only tiles 5 and 6 are black versatiles, and the other versatiles are colored by two colors in order to use them for the border art of the spiral and labyrinth.

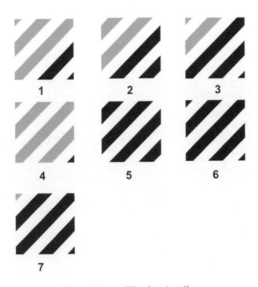

Fig. 9.21.   The basic tiles

Fig. 9.22.   A spiral being created by Slavik Jablan at the Bridges Conference

9.10. Mazes

If the path from entrance to center is not unicursal then we have
a maze [Fri]. For example, a fun house is pictured as a maze in
Fig. 9.23a. Paths from room to room are denoted in the accompanying graph, shown in Fig. 9.23b. You will notice that certain rooms
have access to several other rooms along the pathways.

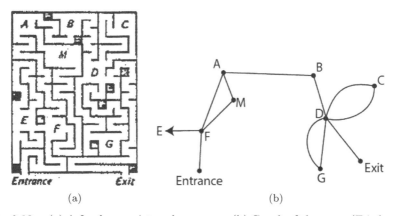

Fig. 9.23.   (a) A fun house pictured as a maze (b) Graph of the maze (Friedman,
2003)

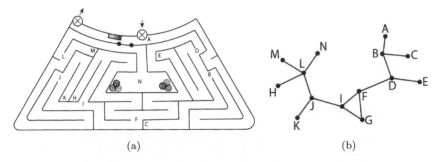

(a)   (b)

Fig. 9.24.   (a,b) The Hampton Court maze and its graph [Friedman, 2003]

Fig. 9.25.   Example of another maze [N. Friedman, 2003]

The Hampton Court maze was commissioned around 1700 by William III. It covers a third of an acre, is trapezoidal in shape, planted with high hedges, and is the oldest surviving hedge maze. Its patterns shown in Fig. 9.24 along with its access graph whose vertices are the points at which the path bifurcates.

I invite you to draw the graph from entrance to center for the maze in Fig. 9.25.

9.11. Conclusion

We have seen how a wide range of mathematics and design has emerged from the concept of a meander. Knots, frieze patterns, labyrinths and mazes all owe their existence to the concept of a meander. Future work in these fields must take this into account.

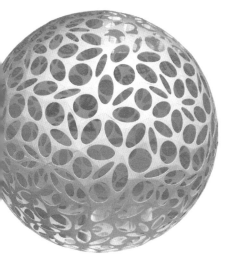

Chapter 10

Why is a Donut Like a Coffee Cup?: An Introduction to Topology

10.1. Introduction

As rich as our everyday experience is, mathematics can present you with things that take you beyond your imagination even in the realm of the elementary.

Take for example the three closed curves shown in Fig. 10.1. The closed curves in Figs. 10.1a and 10.1b are simple because they have no self-intersections whereas the figure eight curve in Fig. 10.1c is closed but not simple. And it is obvious that these *simple* closed curves divide the plane into an inside space and an outside space. This brings us to the *Jordan curve theorem* which says that any simple closed curve divides space into an inside and an outside. As obvious as this is for our examples, consider the closed curve in Fig. 10.2.

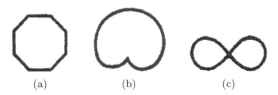

(a)   (b)   (c)

Fig. 10.1.   (a) (b) are simple closed curves, (c) is not simple

Fig. 10.2.   A complex simple closed curve

You can see this result is no longer obvious. Can you think of a way of determining whether any given point is inside or outside?

10.2. Torus and Sphere

In the field of *topology*, you can deform an object without cutting to get an equivalent object. In Fig. 10.3 this is done by deforming a coffee cup until it takes the form of a donut. Therefore, the donut and coffee cup are said to be *topologically equivalent*.

The outer surface of the donut is referred to as a *torus* and takes the form of a rubber tire. Now that we have introduced the *torus*, we can compare it with the outer surface of a *sphere* as shown in Fig. 10.4. You will notice that no amount of deforming without cutting of the torus will ever make it into a sphere. In other words, the torus and sphere are *not topologically equivalent*. And this is born out in their different properties. For example, if you remove a closed curve, or *loopcut,* from the surface of the sphere (see Fig. 10.5a) the sphere is now in two pieces. However, can you see from Fig. 10.5b that a loopcut made on a torus can still leave the torus in one piece. In fact, you can make two loopcuts to the torus, and it still remains in one piece. One cut will result in a cylinder and the other will be

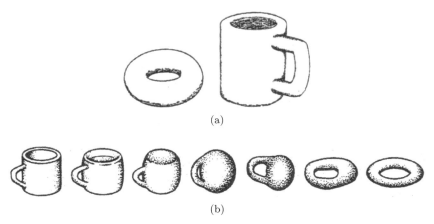

(a)

(b)

Fig. 10.3. (a) is a donut and a coffee cup, (b) is a homeomorphic transformation from a donut to a coffee cup

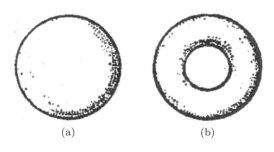

(a)   (b)

Fig. 10.4. (a) Sphere, (b) Torus

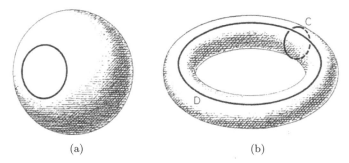

(a)   (b)

Fig. 10.5. (a) One loopcut divides a sphere into two pieces; (b) Two loopcuts keeps the torus in one piece

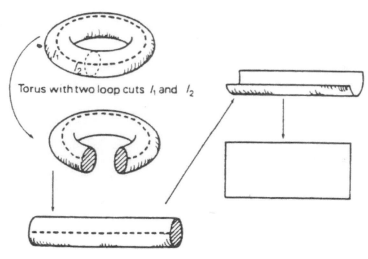

Fig. 10.6.   With two loopcuts the torus opens first to a cylinder and then to a rectangle

a cut from one point on the circular edge of the surface to another point on the edge which opens the cylinder to a rectangle as shown in Fig. 10.6. We say that the sphere has a *Betti number* of 0 whereas the torus' Betti number is 2 since two cuts can be made, yet the torus still remains in one piece.

A map is a collection of faces, edges, and vertices such that each face can be deformed to a point while information is given as to how the vertices are connected by the edges. For example, Fig. 10.7a illustrates a face with two vertices and two edges in which the face can be placed on a rubber sheet and deformed to a point, whereas the shaded area of Fig. 10.7b is not a face since it cannot be deformed to a point without interfering with the inner edge. You can draw a map on a sphere or a balloon with four faces such that each face shares an edge with the other three faces as shown in Fig. 10.8. In doing this bookkeeping, the outside face counts as a fourth face. To see this, take a scissors and cut out faces 1, 2, and 3. In that case face 4 is the sphere-like triangular face that is left over. Yet, you cannot draw a map on the sphere with five faces such that each face shares an edge with the other four (try it). We say then that the sphere has a *chromatic number* of 4 and, as a result, given a map on the

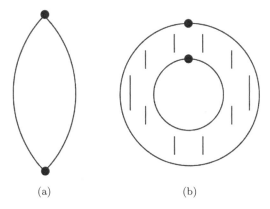

Fig. 10.7.   (a) map, (b) not a map

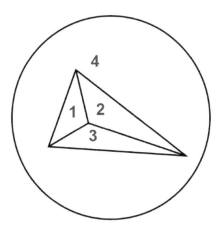

Fig. 10.8.   Map drawn on a sphere. Face 4 is the outside face

sphere, we may need as many as four colors to color it so that no two faces that share an edge have the same color. However, we will never need more than four colors to color a map on the sphere. This was proven by Kenneth Appel and Wolfgang Haken in 1976. On the other hand, along the surface of a torus, you can draw a map with seven faces such that each face shares an edge with the other six, and so some maps drawn on a torus will require seven colors, but never more than seven, Fig. 10.9 shows seven faces drawn on a rectangle. When the rectangle is rolled up first to a cylinder then to a torus

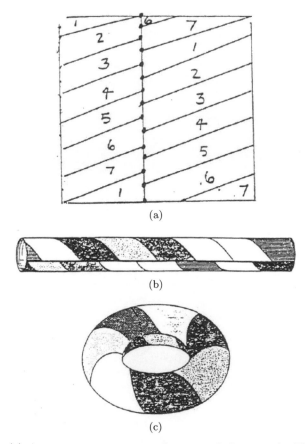

Fig. 10.9. (a) A map on a torus opened to a period rectangle. The map has seven faces with each face bordering the other six; (b) & (c) Another view of the map of a torus opened to a cylinder

each face will share an edge with the other six faces, i.e., the torus has a chromatic number of 7. We will show you another way to create such a 7-colored map on a toroidal polyhedron in Chap. 11.

Exercise 1: Figure 10.6 shows how a torus can be folded up from a rectangle. The torus can then be folded so it is flat. Once you have such a torus, it is instructive to try cutting it in various ways and see if the results surprise you. Try vertical, horizontal, and diagonal cuts.

Remark: The sphere has many things in common with the plane. In fact, there is a one-to-one relation between the points on a sphere and the points on the plane as shown in Fig. 10.10, where any point p in the plane is mapped from a point p' on the sphere except for the North Pole. This mapping is called a *stereographic projection*. The North Pole maps to all of the points at infinity on the plane as shown in Fig 10.10. This transformation of sphere to plane is called a stereographic projection. The plane has the same chromatic number as the sphere as you will see in Exercise 2.

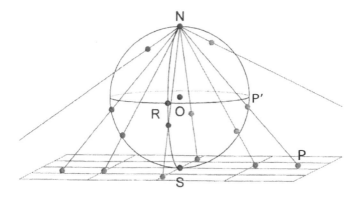

Fig. 10.10.   The stereographic projection

Exercise 2: Draw a map on the plane with four faces such that each face shares an edge with the other three. Then show that you cannot draw a map with five faces such that each face shares an edge with the other four. In other words, the plane has chromatic number 4. Remember that the outside face must be counted.

10.3. The Euler-Poincare Number

The great mathematician Euler (1707–1783) discovered that all maps on surfaces that are topologically equivalent share a simple number relationship, namely,

$$F + V - E = N,$$

where F, V, E stand for the faces, vertices, and edges of the map. For example, the map in Fig. 10.11 has $F = 3$, $V = 5$, and $E = 6$ and therefore, $N = 2$. Again, remember, the outside counts as a face.

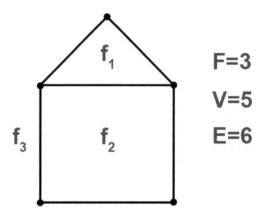

Fig. 10.11.   A map on the plane

Exercise 3: Draw your own maps on the plane, and show that you always get $N = 2$. Try to trick the Euler-Poincare relationship by drawing complex maps.

In Fig. 10.12, you see a picture frame represented as a map on a torus. By counting edges, vertices and faces show that the Euler-Poincare number is $N = 0$. This will be true for all maps on surfaces topologically equivalent to a torus.

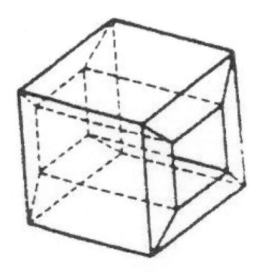

Fig. 10.12.   A map on a torus. Count the faces, vertices and edges

10.4. Mobius Strip

Now let's consider one of the most important and surprising discoveries in mathematics made by the mathematician, August Mobius (1790–1868). However, first we will create a cylinder. We do this by taking the rectangle in Fig. 10.13a and taping two edges together so that point A matches C, and B matches D to form a tube or cylinder. Notice that the cylinder has two circular edges. You can use red paint to paint the outside of cylinder, and then blue to paint the inside. We say that the cylinder is *two sided*. It should be noted that the surface that we are talking about is a mental fiction since, in reality, it has no thickness. Nevertheless, we imagine, as you will see, that it can be painted, twisted, and cut with a scissors.

Now let's create a Mobius strip. Take the rectangle in Fig. 10.13a, but this time give it a 180 deg. twist and identify point A with D and B with C to end up with the Mobius strip in Fig. 10.13b. It will be easier to do this exercise if you use a long rectangular strip. The cylinder clearly has two edges and two sides, an inside and an outside, whereas the Mobius strip has only one edge which you can verify by running your finger along the edge, coming back to where

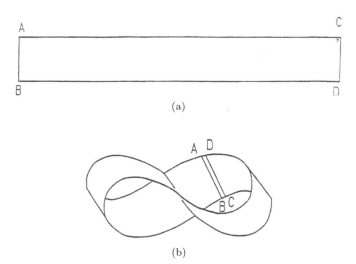

(a)

(b)

Fig. 10.13.   (a) Flat band of paper for making Mobius strip; (b) Assembled Mobius strip you started. If you try painting the Mobius strip, you will find that at the end of your journey both sides of the strip will be painted red, i.e., the Mobius strip is said to be   *one-sided.* Furthermore, you can imagine a sailboat traveling around a Mobius sea with its mast pointed up. When you return to where you started, you will find that the flag now points down. Try to simulate this by moving a matchstick around the center of the Mobius strip. In other words, the Mobius strip cannot be *oriented.* In Fig. 10.14, we see Dr. Mobius represented by a two dimensional human in a Mobius strip universe. If Dr. Mobius travels around the strip, his internal organs will be reversed. The Mobius strip has a Betti number 1 and, as Fig. 10.15a shows, it has chromatic number 6. Note: In Fig. 10.15 the arrows match up, or as we say, are identified after the twist so that the map is continuous across the identified edge. This is known as a *period square.* For example, face 1 borders 2, 3, and 4, while after the twist the two face 1's are identified and now border 5 and 6. Fig. 10.15b shows a periodic square of a torus with seven faces with each face bordering the other six, another example of the seven color problem for a torus. In Fig. 10.15b, opposite edges are identified as are the four corner points.

Exercises 4: Carry out the following three cutting exercises:

a. Cut a Mobius strip down its center. Do you notice that it results in a pair of strips. Determine the nature of these strips. Do they have a single edge? Are they one-sided or two sided?

Fig. 10.14. Dr. Mobius, represented as a two-dimensional human in Mobius strip universe. If Dr. Mobius travels around the strip, his internal organs will be reversed (by C. Pickover [Pic])

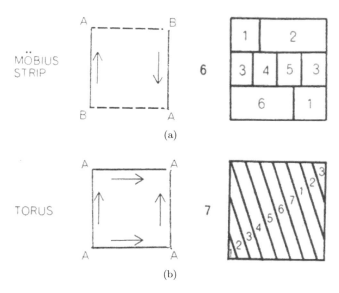

Fig. 10.15. (a) Period map on a Mobius strip; (b) period map on a torus

b. Cut the Mobius strip along an edge 1/3 of the way from one of its outer edges. What did you notice? How many edges does the resulting strip have? Is it one sided or two sided?

c. Now form two cylindrical strips (no twists). Tape them at right angles as shown in Fig. 10.16a and cut them down the center lines. What do you see?

d. Now form two Mobius strips in Fig. 10.16b, by twisting one to the left and the other to the right. Take one of the strips, turn it at right angles to the other and tape the two strips together. Then cut through the center of both strips. Your result should be a pair of hearts. You can use them to express your love for mathematics or send them to someone on Valentine's day.

10.5. Klein Bottle

We now introduce a one-sided surface of interest to topologists called a *Klein bottle*, named after the great 19th century geometer, Felix Klein (1849–1925). He invented what he called a "bottle" whose inside is its outside. However, this bottle exists only in the imagination. We are only able to make models that look like the Klein bottle.

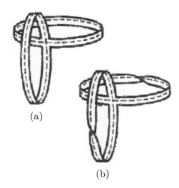

Fig. 10.16. (a) Cut these two cylinders on the dotted lines; (b) Cut these twisted Mobius strips on the dotted lines to obtain a pair of interlocking hearts

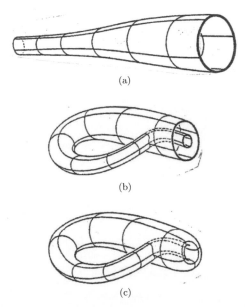

Fig. 10.17. (a) Create a horn (b) Penetrate the wide part of the horn with the narrow part and glue the narrow circle of the horn to the wider circle of the horn (c) The result is a Klein bottle

Klein started with a horn shape tube (see Fig. 10.17a). He thought of putting the smaller part of the tube inside the wider part by making a slit about a quarter of the distance to the end of the tube (see Fig. 10.17b). And then he cemented the open ends of the small end of the tube to the larger end as shown in Fig. 10.17c.

An actual Klein bottle exists only in the 4th dimension, where, in the 3rd dimension, edges only appear to intersect as much as the edges of a cube only appear to intersect in a two-dimensional drawing of the cube. Now, can you tell where the outside ends and the inside begins? If you begin your journey on the "inside" of the bottle, you will soon find yourself outside. This is the mystery of the Klein bottle. It has only one side, like the Mobius band.

To build a Klein bottle from a rectangular piece of paper, follow these steps.

a. Fold the rectangle in half and tape one side of the rectangle to the other side to form a cylinder as shown in Fig. 10.18a.
b. Cut a slit in the cylinder to form a slot about a quarter of the distance to the end of the tube. Be sure to cut only through the part of the tube facing you (see Fig. 10.18b).

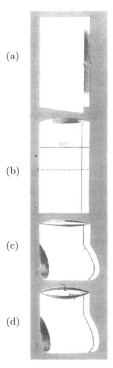

Fig. 10.18.   Folding a Klein bottle from a rectangle (by Martin Gardner [Gar2])

c. Fold the tube into half along dotted line A (see Fig. 10.18b).

d. Insert the bottom of the tube into the slot equivalent to placing the end of the flared tube through the hole above as shown in Fig. 10.18c.

e. Attach the inserted circle around the top of the model (see Fig 10.18d).

Experiment 4: Now that you have the Klein bottle, cut it along the dotted line in Fig. 10.19a. You should end up with a pair of Mobius strip shown in Fig. 10.19b.

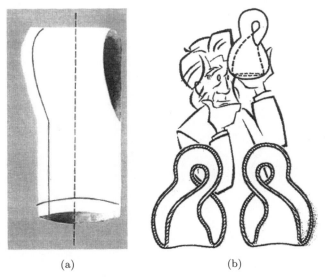

(a)   (b)

Fig. 10.19.   (a, b) A Klein bottle sliced into two halves reveals two Möbius strips

10.6. Mobius Strip and Sculptures

The Mobius strip can now be used to create three 3-dimensional sculptures.

Sculpture 1:

a. In Fig. 10.20, a Mobius strip is built out of squares except for one square where the twist takes place. This enables sculptural forms to be created. Try creating such a Mobius strip.

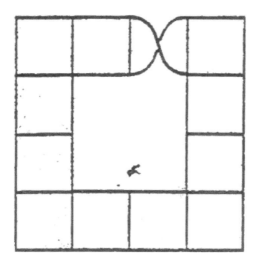

Fig. 10.20. A Mobius strip made out of squares

b. In Fig. 10.21a, the trajectory begins at point P, out of sight and below a square on level 1. The trajectory then moves unseen and emerges into sight on level 2 and travels visibly to an intermediate point, Q.

c. In Fig. 10.21b, the trajectory moves along a visible path descending to level 1, and then back to the original square at point R, but this time above the square.

d. The entire path is shown in Fig. 10.21c. If you cut out the squares corresponding to this trajectory, the result will be a Mobius strip. This is the first Mobius sculpture.

Sculpture 2:

a. Fig. 10.22a shows an exploded view of a single module constructed from squares.

b. Fig. 10.22b shows an exploded view of 4 connected modules.

c. Finally, Fig. 10.22c puts the 4 modules together into a sculpture. All visible tile edges along the perimeter are shown.

Sculpture 3:

a. Fig. 10.23a shows an exploded view of a single module.

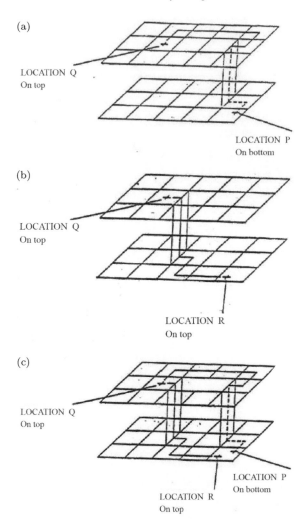

Fig. 10.21.   Steps to Sculpture 1

b. Fig. 10.23b shows an isometric view of 3 modules showing the only 3 allowable orientations for this modular form. All visible tile edges only show edges along the perimeter are shown.

c. Fig. 10.23c shows an exploded view of 4 connected modules Only edges on the perimeter are shown.

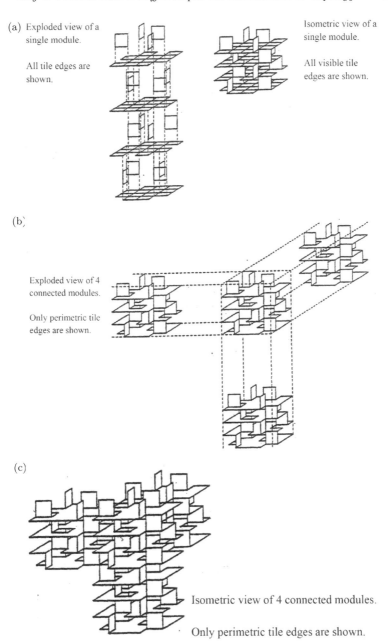

(a) Exploded view of a single module.

All tile edges are shown.

Isometric view of a single module.

All visible tile edges are shown.

(b) Exploded view of 4 connected modules.

Only perimetric tile edges are shown.

(c)

Isometric view of 4 connected modules.

Only perimetric tile edges are shown.

Fig. 10.22.   Steps to Sculpture 2

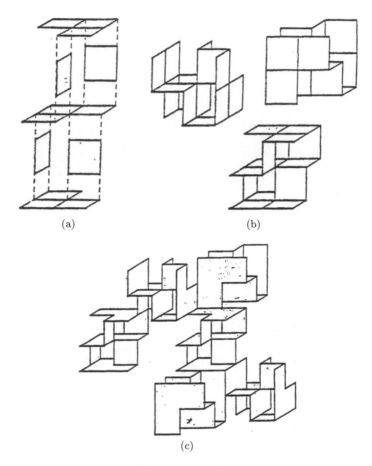

(a)   (b)

(c)

Fig. 10.23.   Steps to Sculpture 3

10.7. Klein Bottle Sculpture

Fig. 10.24a shows a sculptural model of the Klein bottle made in
two parts. Do you see that when the two parts are joined, it forms
a surface with no inside or outside? When you begin coloring the
surface, you will find that only one color is needed. You can construct
this surface by doing paper cutting, or you can fold the model in
Fig. 10.24b up from the plane. It is best to make the parts out of
stiff construction paper. I leave it to the readers, to construct this
Klein bottle sculpture.

(a)

(b)

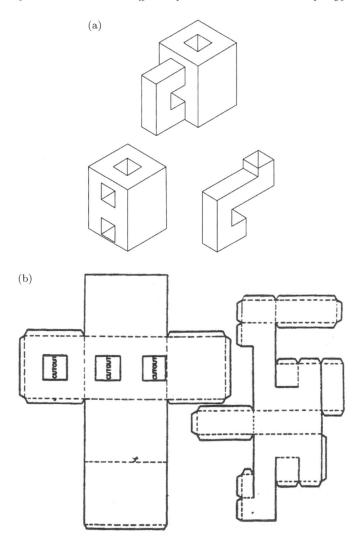

Fig. 10.24.   (a, b) Construction of a Klein bottle

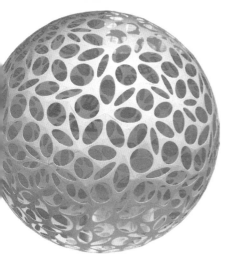

Chapter 11

The Szilassi and Csaszar Polyhedra

11.1. Introduction

The constructions of two toroidal polyhedra related to the 7-color problem of topology are presented in this chapter.

11.2. Construction 1: Szilassi Polyhedron

a. The Szilassi polyhedron, shown in Fig. 11.1, was created by the Hungarian mathematician, Laslo Szilassi. It is an example of a polyhedron topologically equivalent to a torus that has seven faces with each face, a hexagon, sharing an edge with the other six faces. Therefore, the Szilassi polyhedron has a chromatic number of 7, requiring seven colors to color its faces so that two faces sharing an edge have different colors.

b. The assembly of the polyhedron is intended as a puzzle. The 7 faces are shown in Fig. 11.2. Notice that they are paired, with identical faces: 2 = 5, 1 = 6, and 3 = 4. Face 7 is unpaired. Fig. 11.2 is meant to be an aide to the construction.

c. In Fig. 11.3, measurements are given so that the three paired faces and the one unpaired face can be reconstructed to scale. The Szillassi polyhedron can then be reconstructed using Fig. 11.2.

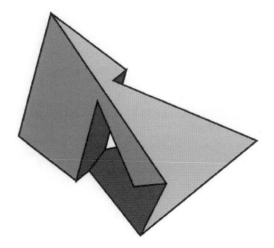

Fig. 11.1.   A Szilassi polyhedron

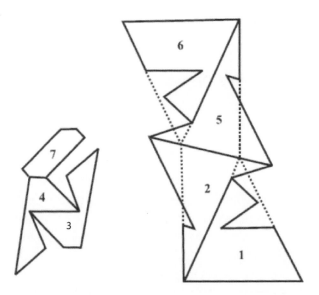

Fig. 11.2.   The seven faces of the Szilassi polyhedron

d. Or, you can choose the size that you would like, photocopy the faces, cut them out, and assemble the polyhedron in this way.

Construct a Szilassi polyhedron.

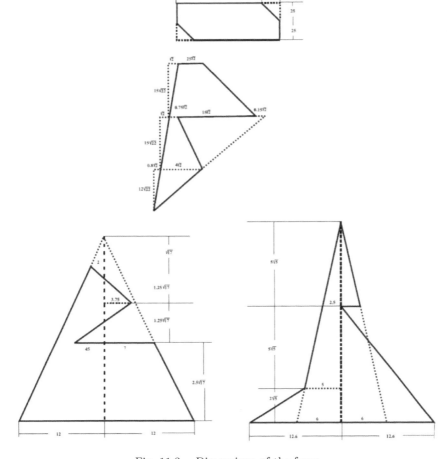

Fig. 11.3.   Dimensions of the faces

11.3. Construction 2: Csaszar Polyhedron

a. In Construction 2, you are asked to recreate the dual to the Szilassi polyhedron known as the Csaszar polyhedron, created by another Hungarian mathematician, Akos Csaszar. As for the dual, this polyhedron has 7 vertices with each vertex connecting to the other 6 while the edges of these duals are paired. Whereas the Szilassi polyhedron has 7 faces, 14 vertices and 21 edges, the Csaszar polyhedron has 7 vertices, 14 faces and 21 edges. The Euler-Poincare number for both the Szilassi and Csaszar polyhedra is $F + V - E = 0$, the number for a torus as we saw in Sec. 10.3.

b. To illustrate the dual relationship between the Szilassi and Csaszar polyhedron, a dot is placed near the center of each edge, as shown in Fig. 11.4, and another dot is placed within the face (the latter point should be chosen carefully to minimize clutter in the next step).

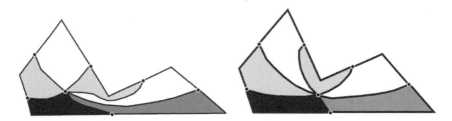

Fig. 11.4.   Csaszer, the polyhedron, dual to the Szilassi

c. Connect the dot in each of the seven faces to each of the 6 dots on the edges surrounding that face. The connecting lines should go through the midpoint of the edge shared by two adjacent faces. The lines may be straight, but some will have to be curved as shown in Fig. 11.4.

d. The Csaszar polyhedron is pictured in Fig. 11.5. It can be constructed directly or by using the pattern shown in Fig. 11.6 where the numbers in both figures represent the vertices. Vertices 2, 5, 3, and 4 were selected to form a regular tetrahedron. The edge lengths and dihedral angles are listed in Table 11.1. Color the faces using the fewest colors. This is equivalent to coloring the vertices of the Szilassi polyhedron so that no two adjacent vertices (connected by an edge) have the same color. Note: A dihedral angle is the angle between two planes in edge view, i.e., when the edge between the two faces appears as a point.

Construct a Csaszar polyhedron.

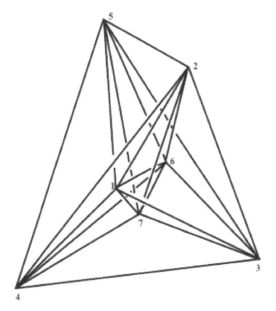

Fig. 11.5. In the dual each of the 7 vertices connect, by an edge, to the other 6 vertices

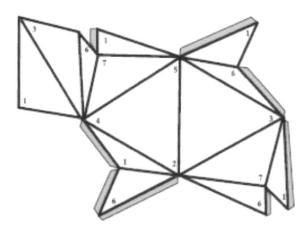

Fig. 11.6. A model of the Csaszar polyhedron I created by folding from the plane

Table 11.1.   Edge Lengths and Dihedral Angles of a Csaszar Polyhedron

Edge	Edge Length	Dihedral angle (degrees)
$(1 - 6)$	10.00	76.13
$(2 - 5)$	24.00	70.53
$(3 - 4)$	24.00	54.43
$(2 - 4) = (5 - 3)$	24.00	51.05
$(2 - 3) = (5 - 4)$	24.00	52.72
$(3 - 7) = (4 - 7)$	12.89	340.13
$(2 - 7) = (5 - 7)$	17.15	74.42
$(1 - 5) = (6 - 2)$	18.69	339.32
$(1 - 2) = (6 - 5)$	12.55	156.85
$(1 - 4) = (6 - 3)$	12.55	204.47
$(1 - 3) = (6 - 4)$	17.36	41.67
$(1 - 7) = (6 - 7)$	5.86	243.50

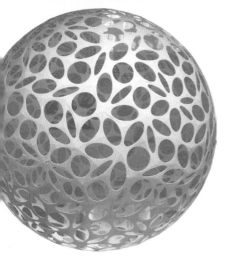

Chapter 12

Curves of Constant Width and a Three Dimensional Sculpture that Rolls

12.1. Curves of Constant Width

Every automobile based on our experience has circular wheels. However, did you know that there are many other wheel shapes besides circular that can function in the capacity of a wheel? In fact, there are an unlimited number of other possibilities that can function as a wheel. In this chapter we will construct several classes of these so-called *curves of constant width* [Gar3].

Note that a convex planar curve is a curve such that any line between two points within the curve lies entirely within the curve such as an ellipse. The *diameter* of a convex closed curve is defined as follows.

Draw a straight line that touches the curve without intersecting it. This can always be done if the curve is convex. Next, bring a parallel line in from outside of the curve until it also touches the curve on the opposite side. The distance between the lines is defined to be a diameter of the space enveloped by the curve. If for every point on the closed curve the diameter is the same then we say that the closed curve has constant width.

a. The simplest non-circular curve of constant width has been named the Reuleaux triangle after Franz Reuleaux (1829–1905),

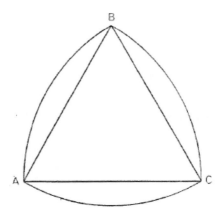

Fig. 12.1.   A Reuleaux curve (by Martin Gardner [Gar2])

an engineer and mathematician. To draw a Reuleaux triangle, begin with an equilateral triangle *ABC* as shown in Fig. 12.1. With compass point on point *A* draw an arc of a circle connecting *B* and *C*. Do the same with points *B* and *C* of the triangle. It is easy to see that this curve has constant width.

Note that the Reuleaux triangle rotates within a square as shown in Fig. 12.2. This led Harry James Watt, an English engineer in 1914 to invent a rotary drill based on the Reuleaux triangle capable of drilling square holes.

b. Next construct a symmetrical rounded-corner curve of constant width. Referring to Fig. 12.3, extend side *AB* of the triangle to *F* and *I*, *BC* to *E* and *H* and *CA* to *D* and *G* where *AD*, *AI*, *BE*, *BF* and *CG*, *CH* are all equal. From points *A*, *B*, and *C* draw arcs, *EF*, *GH*, and *ID*. Finally, from *A*, *B*, and *C* draw arcs: *FG*, *HI*, and *DE*. This completes the rounded corner curve of constant width.

c. Finally, draw any number of intersecting line segments as shown in Fig. 12.4. An asymmetric curve can always be drawn. When a pair of lines intersect, place the compass point on the point of intersection of the lines and draw an arc of a circle. Continue to do the same with every pair of intersection lines and connect the arcs. I will leave the details to the reader. In general, this curve will be asymmetric.

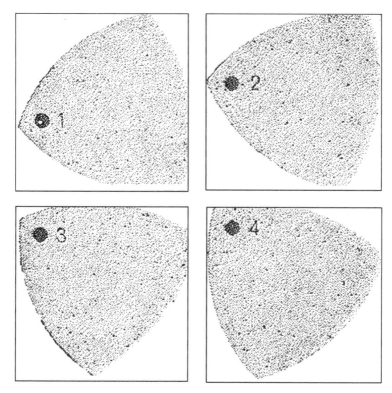

Fig. 12.2.   Rotating a Reuleaux curve results in a square hole (by Martin Gardner [Gar2])

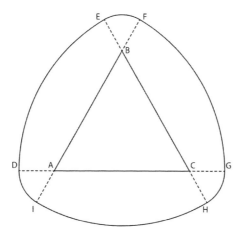

Fig. 12.3.   A Reuleaux curve with curved edges (by Martin Gardner [Gar2])

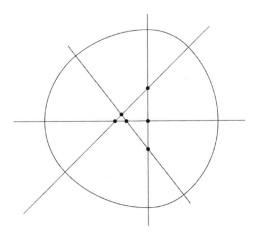

Fig. 12.4.   A Reuleaux curve created from a set of intersecting lines

Exercise:

Construction 1: Construct out of cardboard a Realeaux curve with constant edges and check to see that it rolls.

Construction 2: Construct out of cardboard a Reuleaux curve from a set of intersecting lines and check to see if it roles.

Construction 3: Connect pairs of curves of constant width with an axle and mount these on a toy car to illustrate an automobile with odd shaped curves of constant width.

12.2. Three-dimensional Sculptures that Roll

We have seen in this chapter how structures with shapes related to an equilateral triangle can roll like a wheel. In this section we create a 3-dimensionl object, related to an octahedron, that also rolls.

Dick Esterle, an architect and toy inventor, has created a table-top sculpture based on the edges of an octahedron (see Fig. 12.5). Esterle begins with two circles having squares inscribed in them. A semicircle is removed from each circle, and a third square is added to give rise to 3 flat pieces, as shown in Fig. 12.6a,b,c. Struts form the edges of each of the flat pieces, with slots cut into the vertices so that when the pieces are assembled, they click into place. Enlarged copies

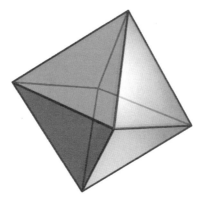

Fig. 12.5. Octahedron

of the 3 flat pieces can better show the corner slots, and these copies can be cut out of stiff cardboard and used to assemble the sculpture. Note that all of the vertices on the 3 flat pieces are closed except a single vertex on one of the squares in the circle which is open. A $3/4$ view of the assembled sculpture is shown in Fig. 12.7 showing the underlying octahedron. Front and top views of the sculpture are shown in Fig. 12.8.

Once you have the completed sculpture, you can see that it rolls in a single direction.

Construction:

Construct Esterle's sculpture and watch it roll. You can construct the 3 elements out of stiff cardboard, as described above, or create it with a 3-D printer.

This sculpture is from a family, of so-called *Oloids* first introduced by the designer Paul Schatz in 1929. Esterle's sculpture is classified as a sphericon. Sphericons were discovered by Colin Roberts, a carpenter from the UK, in 1919 and by David Hirsch who invented a device for generating a meander motion. The device consisted of two perpendicular half-discs joined at their axes of symmetry. Dancer and sculptor Alan Boeding from the MOMIX dance company, in 1979, used a sphericon sculpture in which two crosswise semicircles, each passing through the other, were used to choreograph the dance,

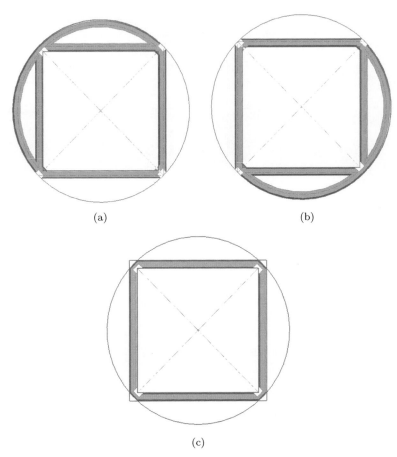

Fig. 12.6. 3 flat pieces for assembly. (a) Flat piece showing the square in a semicircle with an open vertex; (b) A second flat piece showing a square in a semicircle; (c) The third flat piece is a square

"Circle Walker." In 1984 MOMIX choreographed "Dream Catcher" which captured the rolling motion of the Oloid. We will have more to say about Oloids on the website for this book.

Recently, *Platonicons* have been added to this list of meandering objects, and they have a relationship to objects of constant width. Bill Gosper's ambiguous roller, "Lissajous Roller," is a play on these objects in which an optical illusion occurs when it is rolled along its path.

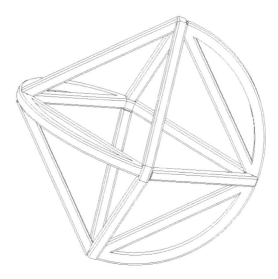

Fig. 12.7.   A view of the rolling sculpture

Fig. 12.8.   Photo of the rolling sculpture

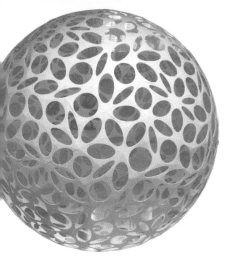

Chapter 13

An Introduction to Fractals

13.1. Introduction

Euclidean geometry has had a major impact on the cultural history of the world. Not only mathematics, but art, architecture, and the natural sciences have utilized the elements of Euclidean geometry or its generalizations to projective and non-Euclidean geometries. However, by its nature, Euclidean geometry is more suitable to describe the ordered aspects of phenomena and the artifacts of civilization rather than as a tool to describe the chaotic forms that occur in nature. For example, the concept of a point, line, and plane which serve as the primary elements of Euclidean geometry, are acceptable as models of the featureless particles of physics, the horizon line of a painting, or the facade of a building. On the other hand, the usual geometries are inadequate to express the geometry of cloud formation, the turbulence of a flowing stream, the pattern of lightning bolts, the branching of trees and alveoli of the lungs, the variability of the stock market, or the configuration of coastlines.

In the early 1950's, the mathematician Benoit Mandelbrot, aware of work done half a century before, rediscovered geometrical structures suitable for describing these irregular sets of points, curves and surfaces from the natural world. He coined the word fractal for these entities and invented a new branch of mathematics to deal with them. The key to understanding fractals and their applications to the

natural world lie in their embodiment of self-similarity, the endpoint
of an infinite process. Unlike calculus, which was invented by New-
ton and Leibnitz in the 17th century to deal with the variability and
change observed in dynamic systems such as the motion of the planets
and mechanical devices using curves that were mostly smooth, fractals are generally depicted by structures that are nowhere smooth.
In fact, many of the so-called pathological curves which were discovered by mathematicians studying the foundations of calculus and
then banished from its further study, became the starting points for
this new discipline of fractals.

Let us see how fractals are generated [Cra], [Kap1].

13.2. The Koch Snowflake

The Koch snowflake is created by the following steps:

1. Begin with a line segment as shown in Fig. 13.1a.
2. Subdivide the line segment into three equal parts and replace the
   middle third by a pair of edges of the same length to form a
   structure with four equal line segments as shown in Fig. 13.1b.
3. Apply steps 1 and 2 to each of the resulting four line segments
   (see Fig. 13.1c).
4. Continue this process to later stages in the development of the
   Koch snowflake. A later stage is shown in Fig. 13.1d.
5. The same process could also be carried out beginning with an
   equilateral triangle as shown in Fig. 13.1.
6. Starting with an equilateral triangle, snowflake patterns are
   obtained as shown on the right side of Fig. 13.1.

It turns out that at the end of this infinite process, the snowflake
will be infinite in length and nowhere smooth. In fact, the distance,
as measured along the snowflake curve between any two points on
the snowflake, no matter how seemingly "close", will be infinite. Also
observe in Fig. 13.1d that each segment of the fractal appears like the
whole except at a smaller scale. This self-similarity becomes exact at
the infinite stage. We will find this self-similarity to be ubiquitous to
fractals.

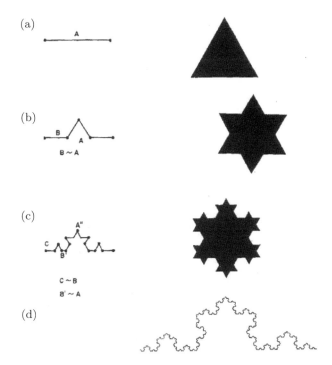

Fig. 13.1. Three steps in the evolution of the Koch curve

13.3. A Fractal Tree

1. Begin with a rudimentary tree branch with five branch tips (see Fig. 13.2a).
2. Replace each branch tip by an exact miniature of the original branch (see Fig. 13.2b). The tree now contains 25 branches.
3. Replace each of the 25 branches again with a miniature of the original so that Fig. 13.2c possesses 125 branches.
4. Repeat this process infinitely to obtain the fractal tree branch shown in Fig. 13.2d.

Remark: In the early stages of this process, the tree branch is not self-similar. However, in advanced stages its self-similarity begins to become evident, and it is fully manifested in the infinite stage of its development. Of course, we can never reach this infinite step, and so we must stop the process at a sufficiently late stage.

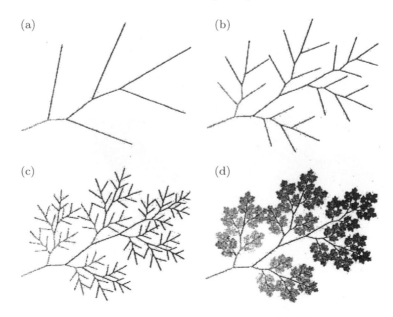

Fig. 13.2.   Evolution of a fractal tree

13.4. A Moonscape

Annalisa Crannell and Marc Frantz [Cra] put forth the following fractal simulation of a moonscape.

1. Begin with a circular "crater" (see Fig. 13.3a) placed in a square.
2. Randomly add to the square eight circular "craters" of width 1/3 the original (see Fig. 13.3b).
3. Repeat this process again with the addition of 64 craters of size 1/9 (see Fig. 13.3c).
4. In the following two stages, 512 and 4096 craters are added as shown in Figs. 13.3d and 13.3e.
5. A photograph of the Moon is shown in Fig.13.3f.

Remark: Each segment of the square looks very much like every other segment, again, reflecting the self-similarity of the moonscape. After all, outside influences that led to the moonscape, such as meteor

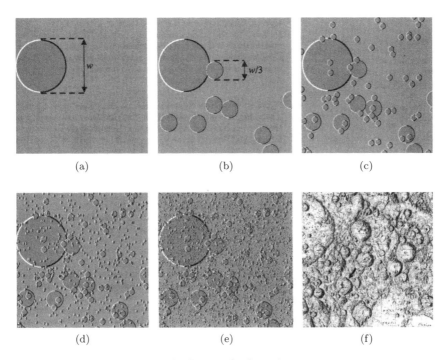

(a)   (b)   (c)

(d)   (e)   (f)

Fig. 13.3.   Evolution of a fractal moonscape

impacts, act in a similar way within each sub-portion of the square
and should, therefore, result in a scape with similar appearance.

13.5.   A Cauliflower

Purchase a head of cauliflower from your local supermarket. Observe
how each of the florets appears to be self-similar to the whole
cauliflower. We can artificially reproduce a facsimile of the cauliflower
by the process shown in Fig.13.4, which needs no further explanation [Cra].

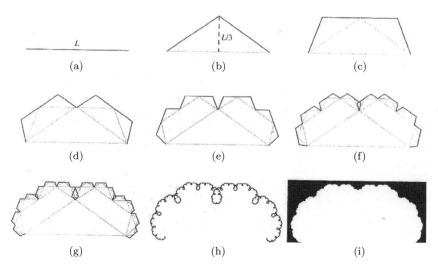

Fig. 13.4.   Eight steps in the evolution of a fractal cauliflower

13.6. The Dragon Curve

The Dragon curve [Fram] is a striking example of the evolution of a fractal based on $\sqrt{2}$ as shown in five stages in Fig. 13.5. Beginning with step 0 each succeeding stage of the Dragon curve is represented by solid lines. At each stage the previous stage is represented by a dotted line and leads to the derivation of the next stage after a sequence of right-left-right-left... twists and turns. After an infinite number of stages the final Dragon curve emerges. I recommend that you follow the sequence of stages leading to stage 4 to derive stage 5.

It is also evident from Fig. 13.5 that at each stage there are two copies of the previous stage shown in red and blue. As the curve evolves, it turns out that it can be depicted by: 1, 2, 4, 8, ... self-similar copies. The curve at a late stage is shown in Fig. 13.6 in four colors.

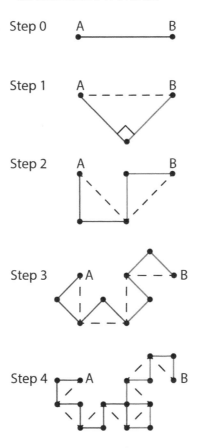

Fig. 13.5.   Three steps in the evolution of the Dragon curve in two colors

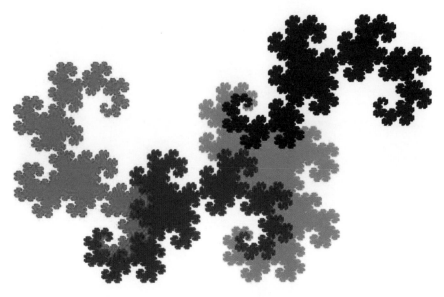

Fig. 13.6.   A late stage in the evolution of the Dragon Curve in four colors

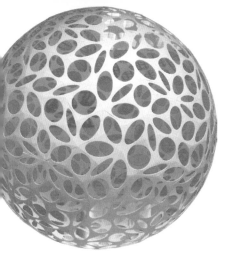

Chapter 14

Creating a Fractal Wallhanging

14.1. Isometries

We wish to create a fractal pattern beginning with an undifferentiated black square. However, we must first understand some things about the elements of 2-dimensional symmetry. This involves a class of transformations known as isometries.

A transformation is a 2-dimensional *isometry* if it transforms a region of the plane such that the distance between points is preserved. It can be shown that there are four types of transformations that do this: translations, T, rotations, S, reflections, R, and glide reflections, G. There is one additional transformation, the identity, I, which transforms each point to itself.

a. A *translation*, denoted by T_V, is characterized by a vector $\vec{v}$. To carry out the translation, place the tail of the vector at a point, p, and its transform, p', appears at the tip, shown in Fig. 14.1a with multiple translations in Fig. 14.1b.

b. A *rotation* S_θ^O is characterized by a point O and an angle θ. Each point of the plane transforms by rotation about O counterclockwise through angle θ about the given point O as shown in Fig. 14.2a with multiple rotations shown in Fig. 14.2b.

c. To characterize a *reflection* R_M you need to designate a line M in the plane referred to as a mirror line. Given point p, its transform

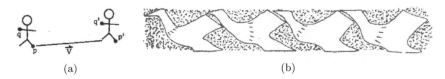

(a)   (b)

Fig. 14.1.   (a) Representation of a translation; (b) Multiple translations

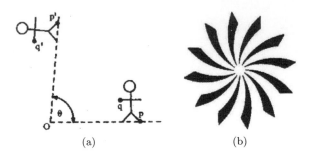

(a)   (b)

Fig. 14.2.   (a) Representation of a rotation; (b) Multiple rotations

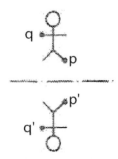

Fig. 14.3.   Representation of a reflection

p' can be found on the other side of the mirror line an equal distance to the mirror as shown in Fig. 14.3.

d. The glide reflection $G_M^{\vec{v}}$ is less familiar. It is characterized by a line M in the plane referred to as a glide line and a vector $\vec{v}$ parallel to the glide line. Given a point p, simply transform p by vector $\vec{v}$ parallel to the glide line to p'', and then reflect it to p' as if the glide line were a mirror line as shown in Fig. 14.4a. If this is done repeatedly, as in Fig. 14.4b, it results in the image of footsteps in the snow.

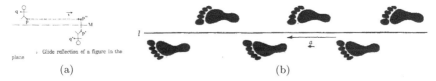

Fig. 14.4. (a) Representation of a glide reflection; (b) Multiple glide reflections (footprints in the snow)

14.2. The Symmetry Group of the Square

Various classes of symmetry can be expressed by determining the set of isometries that keep a particular region of the plane invariant. By this we mean that the points may move to new positions but the region remains the same. The set of transformations that do this are referred to as the symmetries of that region. Take, for example, the symmetries of a square. We can list all of the isometries that transform the points of the square but leave the square unchanged as follows.

If the center of the square is taken to be the point of rotation O, then rotations of 90 deg., 180 deg., 270 deg. leave the square invariant, i.e., I, S_{90}, S_{180}, S_{270}. A rotation of 360 deg. about O brings the square back to its origin and so must be the identity, I.

Reflections in vertical and horizontal lines through the center of the square are symbolized by V and H, while reflections about the diagonals of the square, symbolized by $D_{\div}$ and D_{-}, complete the list of isometries that leave the square invariant. These eight transformations: $\{I, S_{90}, S_{180}, S_{270}, V, H, D_{\div}, D_{-}\}$ make up what mathematicians refer to as the symmetry group of the square. These symmetries are illustrated in Fig. 14.5.

14.3. Creation of a Fractal by the Iterative Function System

Fractals are objects that are self-similar at a variety of scales. By self-similar, we mean that the fractal contains copies of itself transformed by the set of symmetries at a variety of scales from large to infinitesimal. To create a fractal by the *Iterative Function System* (IFS) [Cra], begin with an arbitrary design drawn within a 1 unit × 1

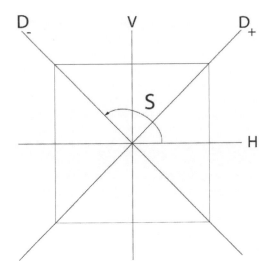

Fig. 14.5.   Isometries keeping a square invariant

unit square and iteratively subject this design to a given set of transformations. Four fractals generated by IFS are shown in Fig. 14.6. We next show how to generate the fractal pictured in Fig. 14.6d.

a. Begin with a 1×1 square colored black so that the starting design is a black square as shown in Fig. 14.7a.

b. Reduce the black square to half its size and,

    i. Transform it by the identity, I, and place it in the lower right-hand quarter of another 1×1 square.

    ii. Rotate it 90 deg. counterclockwise, S_{90}, and place the result in the upper left-hand quarter of the 1×1 square.

    iii. Rotate it by a half-turn, S_{180}, and place it in the lower left-hand quarter of the 1×1 square.

    iv. Leave the upper-right hand quarter empty as shown in Fig. 14.7b. This results in an L-shaped pattern.

c. Reduce the L-shaped pattern within the 1×1 square from step b to half its size and repeat the three transformations given in step b to get the pattern in Fig. 14.7c.

d. The next three iterations are shown in Fig. 14.7d, e, f.

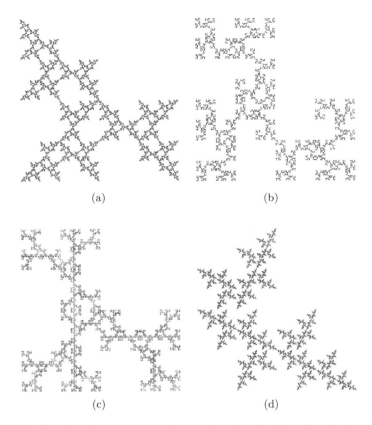

(a)   (b)

(c)   (d)

Fig. 14.6. Four fractals constructed b any the Iterative Function System (IFS)

e. If the three transformations of step b are iterated infinitely, the result will be the fractal pattern shown in Fig. 14.6d. Of course, the iterations must be finitely terminated, as they are in Fig. 14.7f after five iterations, resulting in an approximation to the fractal.

Remark 1: Each iteration reduces the size of the original square by a factor of $1/2$. After several iterations the initial design within the square will be reduced to the size of a pixel. Since any design within the square is eventually reduced to the size of a pixel, you will end up with the same result regardless of the design that you start with.

Remark 2: If you take the final result of this process, shown in Fig. 14.6d, and rotate it by 90 deg. counterclockwise or by a half-turn you will notice that the pattern is identical to the patterns in

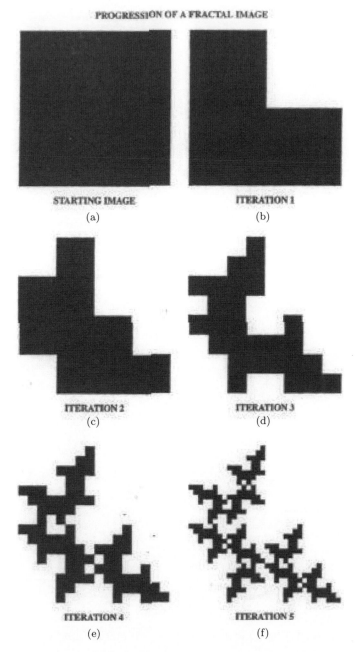

Fig. 14.7.   (a–f) Evolution of the Fig. 14.6d fractal

the upper left-hand and lower left-hand quadrants of the original. This self-similarity is a defining property of all fractals.

Since this fractal was generated by the transformations (I, S_{90}, S_{180}) I will refer to this sequence of transformations as the *genetic code* of the fractal.

14.4. A Fractal Wallhanging

In the genesis of the above fractal the pattern is continuously scaled to half its size so that miniature copies of the final fractal appear at smaller and smaller scales within the whole fractal. Now that you have approximated the fractal, you can recreate it at large scale as a wallhanging as follows [Kap1]:

Materials: Scissors, a stack of sheets of 2 in. × 2 in. fractals of the type shown in Fig. 14.6d, a stack of $8^1/_2$ in. × 11 in. sheets of paper, scotch tape, glue stick, a large 3ft × 3 ft or 5 ft × 5 ft chipboard.

a. Cut out 64 copies of 8 in. × 8 in. squares from the $8^1/_2$ in. × 11 in. sheets of paper.
b. Cut out three 2 in. × 2 in. copies of the fractal in Fig. 14.6d.

  i. On the table place the first copy into the lower right-hand quarter of a 4 × 4 square.
  ii. Rotate the second copy 90 deg. counterclockwise and place it in the upper left-hand quarter of the 4 × 4 square.
  iii. Rotate the third copy by a half-turn and place it in the lower-left hand quarter of the 4 × 4 square.
  iv. Leave the upper right-hand quarter of the 4 × 4 square empty.

c. Create two more copies of this 4 × 4 square.
d. Apply the same three transformations as in part b to the three 4 × 4 squares from step c to get a fractal pattern within an 8 × 8 square. You will need nine copies of the original 2 × 2 fractal to carry this out.
e. Attach with glue stick the 9 copies of the fractal from step d onto one of the 8 in. × 8 in. square sheets of paper.

f. Repeat steps b and c two more times using a total of 27 copies of the 2×2 fractal. This step can be greatly eased if you have access to a Xerox machine.

g. Apply the three transformations to three copies of the 8×8 square from step f resulting in a 16×16 square requiring 27 copies of the original fractal. Tape these three 8×8 squares together on their backside along with a blank square in the upper right-hand quadrant to form an enlargement of the original 2 in. $\times$ 2 in. fractal to one that is now 16 in. $\times$ 16 in..

h. Create two additional copies of the 16 in. $\times$ 16 in. fractal.

i. With three copies of the fractal from steps h move up one more level and create a 32×32 fractal.

j. At this point you can attach this 32×32 square to a 3 ft. $\times$ 3 ft. chipboard. But if you are ambitions and still have enough energy, you can go up one final step to create a 64×64 fractal image.

Since the transformations to create this fractal were all rotations or the identity transformation, the transformations required to create the fractal can be carried out by movements of the original fractal in the plane. This would not have been possible if one or more of the transformations was a reflection. The creation of such fractals are more difficult and must use a computer. We will not address them in this workshop. However, we will show in the next section how to determine the transformations necessary to generate such a fractal, i.e., its genetic code.

14.5. Finding the Code

The possible transformations employed by the IFS system are the eight *symmetries of a square*. These are the distance preserving transformations that keep the square invariant: counterclockwise rotations of 90 deg., 180 deg., 270 deg.: S_{90}, S_{180}, S_{270}; reflections in the vertical, horizontal, and positively and negatively leaning diagonals: V, H, D_+, D_- and the identity transformation, I, as shown in Fig. 14.5. These transformations are all *isometries*. Any one of the three transformations can be one of these eight isometries so that there are $8 \times 8 \times 8$ possible genetic codes. To aid in the process of

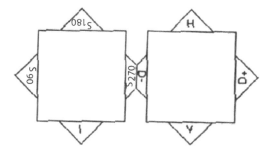

Fig. 14.8. Symmetry finder

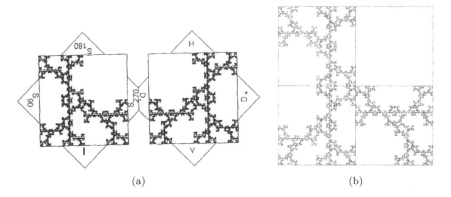

(a)   (b)

Fig. 14.9. Symmetry finder applied to Fig. 14.6c

finding the genetic code of an IFS fractal, use the *symmetry finder* shown in Fig. 14.8. To use this finder, glue the fractal pattern to the left panel of the symmetry finder and its mirror image to the right panel. Cut it out and fold it in half so that the fractal is on the front and its mirror image is on the back. Each time you carry out a transformation by one of the symmetries of a square, record that isometry on the tab at the bottom of the finder as shown in Fig. 14.8. For example, the pattern on the left panel is labeled *I* for the identity, while its mirror image is labeled *V*. This has been done for the fractal in Fig. 14.6c. The fractal and its mirror image are attached to the finder in Fig. 14.9a. Subdivide the fractal into three sectors with a blank in the upper right-hand sector as shown in Fig. 14.9b. Manipulate the finder until you see the pattern in each quadrant of the original fractal emerge on the finder. The transformation will be listed on the tab at the bottom of the finder. Check to see that the genetic code for this fractal is (D_-, S_{180}, V). Now one can reconstruct a scale model in three of the quadrants of a square using the method in Sec. 14.4 by subjecting the fractal to the transformations from the genetic code.

Problem: Find the genetic code for the fractal in Fig. 14.6b using the symmetry finder in Figs. 14.10a and b.

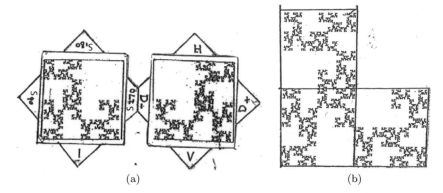

(a)   (b)

Fig. 14.10.   Symmetry finder applied to Fig. 14.6b

14.6. Conclusion

In this workshop we have seen both how to generate fractals by the Iterative Function System beginning with a single black square, enlarge the fractal to create a wallhanging, and find the genetic code of a fractal using the symmetry finder.

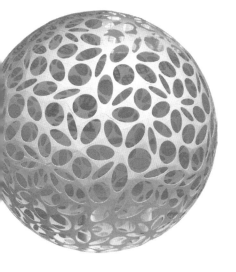

Chapter 15

The Logarithmic Spiral in Geometry, Nature, Architecture, Design, and Music

15.1. Introduction

The *logarithmic spiral* is a mathematical structure with applications to the natural world, spiral galaxies, architecture, design, and music. Yet all of this beauty and structure evolves naturally from a right triangle. This chapter is devoted to exploring the logarithmic spiral and its ramifications.

15.2. Similarity

Two figures are *similar* if corresponding lengths are in proportion. The constant of proportionality is referred to as the *magnification factor*. The *logarithmic spiral* is the only smooth curve that is self-similar. Other self-similar curves are nowhere smooth and have a fractal nature. Arcs on a spiral of angle theta, with respect to the center of the spiral, are similar, i.e., with a magnification or contraction they can be made to lie atop each other. Because of this self-similarity, many forms in nature such as sea shells, sea animals and the horns of horned animals have spiral forms as do spiral galaxies (see Fig. 15.1).

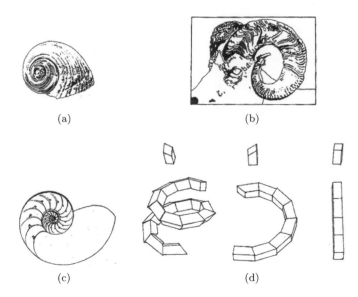

(a) (b)

(c) (d)

Fig. 15.1. Spiral forms in nature

15.3. Construction 1: The Four Turtle Problem

Figure 15.2 shows four turtles, Abner (A), Bertha (B), Charles (C) and Delilah (D). They start at the corners of a square and move towards each other a small distance in a straight line at constant speed with A chasing B, B chasing C, C chasing D, and D chasing A. After the first move, the turtles are in new positions and again move the same small distance towards each other. If this movement continues, the turtles will eventually meet at the corners of a square near the center approximating logarithmic spiral paths. As the distance of each movement shrinks to zero, the paths approach exact logarithmic spirals and the turtles meet exactly in the center [Gar1]. Construct the trajectories of the four turtles.

Remark 1: At all times in this pursuit the four turtles occupy the vertices of a square.

Remark 2: The same set of movements can be carried out with any number of turtles occupying the vertices of a regular polygon. Try it for three turtles.

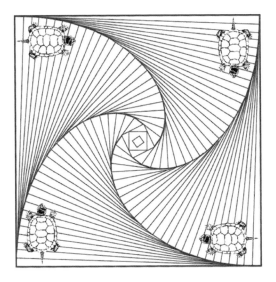

Fig. 15.2. Four turtle problem

15.4. The Self-similarity of a Right Triangle

What I call surgery on a right triangle reveals the self-similarity of
a right triangle (Kappraff, 2015). Three right triangles 1, 2, 3 are
created from a rectangle as shown in Fig. 15.3a. If the right triangles
are cut out and their right angles placed together, as in Fig. 15.3b,
the self-similarity becomes evident, and the lengths labeled a, b, c
satisfy Eq. 15.1 as shown in Fig. 15.3c which is a consequence of the
self-similarity. Eq. 15.1 is known as the *law of the mean proportional*
and was one of two great treasures of antiquity, the other being the
Pythagorean Theorem.

$$\frac{a}{b} = \frac{b}{c} \tag{15.1}$$

15.5. From Right Triangle to Logarithmic Spiral

How does the right triangle lead to the log spiral [Kap1]? In an x, y-
coordinate system, construct a spiral form of what I call vertex points
as shown in Fig. 15.4c. This spiral form is a sequence of right angles
similar to the right angles in Fig. 15.3 that led to Eq. 15.1. I will now
determine the values of the vertex points.

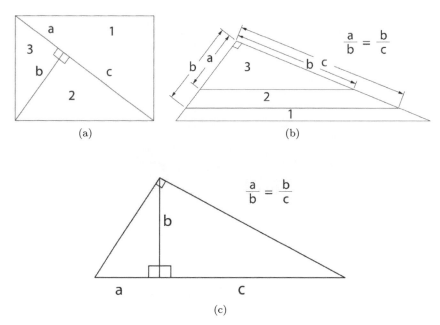

Fig. 15.3.   (a) Surgery on a right triangle; (b) Self-Similarity of a right triangle; (c) The law of the mean proportional considered to be a mathematical treasure of antiquity

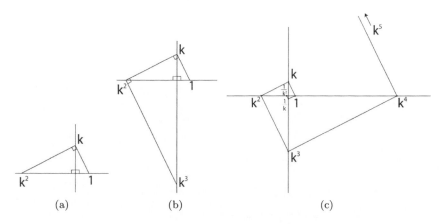

Fig. 15.4.   Vertex points on a log spiral

The first triangle, shown in Fig. 15.4a, has base 1 unit and altitude k units. From this triangle all others follow. Next, construct a second triangle shown in Fig. 15.4a, again, with the use of Eq. 15.1 where,

$$\frac{1}{k} = \frac{k}{x} \quad \text{or} \quad x = k^2.$$

In Fig. 15.4b, notice that triangle 2 and 3 have the same configuration as Fig. 15.3. Therefore, again, from Eq. 15.1,

$$\frac{k}{k^2} = \frac{k^2}{x} \quad \text{or} \quad x = k^3.$$

Continuing in this manner we see that we are generating a sequence of *vertex points* labeled $1, k, k^2, \dots$ of an outwardly spiraling form known as a *logarithmic spiral.* The vertex points are marked by the *radii* of the spiral from the center at the origin.

Note that we can also spiral inwardly where, again, from Eq. 15.1,

$$\frac{k}{1} = \frac{1}{x} \quad \text{or} \quad x = \frac{1}{k}$$

as shown in Fig. 15.4c. Continuing in this way, we find that the radii of the vertex points form a double geometric sequence (Seq. 15.2a) of the radii of both inward and outward logarithmic spirals, as shown in Fig. 15.4c:

$$\dots \frac{1}{k^3} \quad \frac{1}{k^2} \quad \frac{1}{k} \quad 1 \quad k \quad k^2 \quad k^3 \quad k^4 \dots \tag{15.2a}$$

We set $k = 2$ to get the double geometric sequence based on 2,

$$\dots \frac{1}{2^2} \quad \frac{1}{2} \quad 1 \quad 2 \quad 2^2 \quad 2^3 \quad 2^4 \dots \quad \text{or}$$

$$\dots \frac{1}{4} \quad \frac{1}{2} \quad 1 \quad 2 \quad 4 \quad 8 \quad 16 \dots \tag{15.2b}$$

Remark: A sequence is geometric if the ratio of terms is a constant.

In what follows we refer, with all generality, to the geometric sequence with $k = 2$. Consider three successive terms: abc from a geometric sequence. Eq. 15.1 can be rewritten:

$$b^2 = ac \quad \text{or} \quad b = \sqrt{ac} \tag{15.3}$$

where the middle number, b, is referred to as the geometric mean of a and c e.g., for the geometric sequence based on $k = 2$,

$$8^2 = 4 \times 16 \text{ or } 8 = \sqrt{4 \times 16} \quad \text{and} \quad 4^2 = 2 \times 8 \text{ or } 4 = \sqrt{2 \times 8}.$$

The vertex points of the double spiral in Fig. 15.4c are expressed in a polar coordinate system, (θ, r), where θ is a ray from the origin and r is the radius of the vertex point. Positive angles are measured counterclockwise from the positive x-axis while negative angles are measured clockwise. The sequence of vertex points are,

$$\ldots \nu_{-2}, \nu_{-1}, \nu_0, \nu_1, \nu_2, \nu_3 \ldots$$

where vertex point $\nu_0 = (0 \deg., 1)$ is in the direction of the 0 deg. ray, with radius, $r = 1$, while $\nu_1 = (90 \deg., 2)$ is in the direction of the 90 deg. ray with radius $r = 2$, while vertex point $\nu_{-1} = (-90 \deg., 1/2)$ is in the direction of the -90 deg. ray with $r = \frac{1}{2}$.

Proceeding in this manner, all the vertex points of the log spiral are exhibited in Table 15.1 where radius is correlated with angle. Note

Table 15.1.   Radius vs. Angle for vertex points on a log spiral

Vertex Points	$n = \dfrac{\theta}{90}$	r
ν_{-2}	-2	$\dfrac{1}{4}$
ν_{-1}	-1	$\dfrac{1}{2}$
ν_0	0	1
ν_1	1	2
ν_2	2	4
ν_3	3	8
ν_4	4	16
ν_5	5	32
ν_6	6	64
	x	2^x
	$\log_2 y$	y

that in Table 15.1, angle is expressed as multiples of 90 deg. Reading Table 15.1 from left to right results in the Exponential Function,

$$y = 2^x, \tag{15.4a}$$

where x is angle and y is radius. Reading the table form right to left results in the Logarithm Function

$$x = \log_2 y. \tag{15.4b}$$

Appendix 15.1 describes the properties of the logarithm. One important property states that multiplying (dividing) a pair of numbers corresponds to adding (subtracting) their logarithms, e.g., $4 \times 8 = 32$ whereas $2 + 3 = 5$, and $32/8 = 4$, whereas $5 - 3 + 2$. These properties were used to multiply and divide numbers with many digits before the advent of computers by using huge books of logarithm tables. Can you see how this might have been done?

From Table 15.1, observe the defining property of the log spiral.

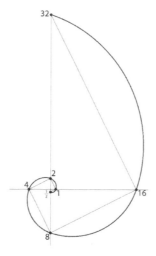

Fig. 15.5.   How do you construct the other points on the log spiral?

Defining Property: If you double the angle, you square the radius.

For example, when the angle doubles from 1 to 2 (or 90 deg. to 180 deg.), the radius squares from 2 to 4 units. If the angle doubles from 2 to 4 (or 180 deg to 360 deg.), the radius squares from 4 to 16 units, etc. You can also observe the defining property in the spiral of Fig. 15.5. A consequence of the Defining Property is that while the angle in Table 15.1 forms an arithmetical progression, the radius forms a geometric progression.

A log spiral spanning the vertex points is shown in Fig. 15.6, How can you find the non-vertex points on this spiral? This can be done analytically by using the Exponential Function, Eq. 15.4a. First of all, we interpolate between successive vertex points, e.g., at angles:

$$\frac{\theta}{90} = \dots \frac{1}{2}, \frac{3}{2}, \frac{5}{2}, \frac{7}{2}, \dots \qquad (15.5a)$$

and use the Exponential Function at these points,

$$r = \dots 2^{\frac{1}{2}} = \sqrt{2}, 2^{\frac{3}{2}} = \sqrt{8}, 2^{\frac{5}{2}} = \sqrt{32}, 2^{\frac{7}{2}} = \sqrt{128}. \qquad (15.5b)$$

We can now add these values to Table 15.1 as new vertex points and begin the process over. For example, we can find the value of the vertex point midway between the ray at $\theta = 0$ and $\theta = 1/4$ (22.5 deg.). From the Exponential Function, it follows that $r = 2^{\frac{1}{4}} = \sqrt{\sqrt{2}}$. If we continue in this way, we can compute the polar coordinates of a dense set of points on the spiral.

We can also determine this dense set of points by compass and straightedge construction of the geometric mean between a pair of vertex points. (See the compass and straightedge construction at the end of this section in Construction 2.)

Given two points a and c, we can use Eq. 15.3 to find the geometric mean, b, between a and c by compass and straightedge construction. e.g., if $a = 1$ and $c = 2$, as they are for the radii of ν_0 and $\nu_1, r = \sqrt{1 \times 2} = \sqrt{2}$. Likewise, if $a = 2$ and $c = 4$, as they are for radii of ν_1, and $\nu_2, r = \sqrt{2 \times 4} = \sqrt{8}$, etc. These values of new radii were computed directly by the analytic method in Seq. 15.5a and b. That the radii are located at the midpoint of successive vertex angles is the result of the Defining Property.

In summary, we have seen that beginning with a single positive number, k, a dense set of points on a logarithmic spiral can be constructed with compass and straightedge.

Construction 2: Construction with compass and straightedge of the geometric mean, $\sqrt{ab}$, between length $a = $ AO and $b = $ BO.

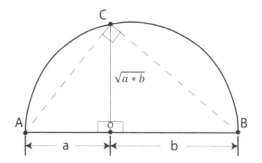

Fig. 15.6.   Construction of $\sqrt{a*b}$, the geometric of a and b

a. On line segment AB in Fig. 15.6, mark off OA $= a$ and OB $= b$ units.
b. Draw a semicircle with AB as diameter, i.e., with $\frac{a+b}{2}$ as radius.
c. Construct the perpendicular line to diameter AB through O, meeting the semicircle at C.
d. It can be shown that angle ACB is a right angle (see Sec. 3.5). Therefore, by Fig. 15.4c and Eq. 15.1, if OC $= x$, then,

$$\frac{a}{x} = \frac{x}{b} \quad \text{or} \quad x^2 = ab \quad \text{or} \quad x = \sqrt{ab} \quad \text{QED.}$$

Problem: Use Construction 2 to find the radii that are geometric means between the pair of radii: $(1, 2)$, $(2, 4)$, $(4, 8)$, $(8, 16)$ and add these radii to the vertex points on the log spiral. Remember that the geometric mean is placed at an angle halfway between successive rays.

15.6. Dynamic Symmetry

The logarithmic spiral is also intrinsic to a construction used to replicate proportions of a rectangle known as the *Law of Repetition of Ratios*. It was used during the classical era in ancient Greek and Roman architecture and design and later in the Renaissance. Its importance was recognized by Jay Hambridge who coined the term *Dynamic Symmetry* to describe this process [Ham], [Edw], [Kap1].

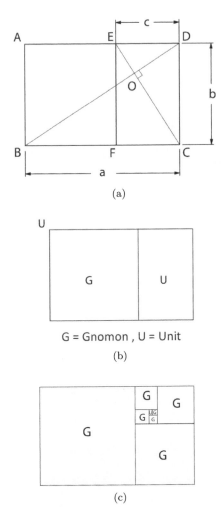

Fig. 15.7. Dynamic symmetry: (a) Construction; (b) Gnomon and Units; (c) Whirling Gnomons

Begin with some geometric form or pattern which we call a *unit*, and add another form or pattern called a *gnomon G* which enlarges the unit *U* while preserving its proportions. In this sense, consider the unit to be a rectangle *ABCD* with sides in the ratio *a* : *b* and draw a line segment *EC* from one vertex that intersects at right angles at *O*, a diagonal *DB* of the rectangle as shown in Fig. 15.9a. In this way the rectangle is divided into two rectangles one of which is a unit *CDEF*

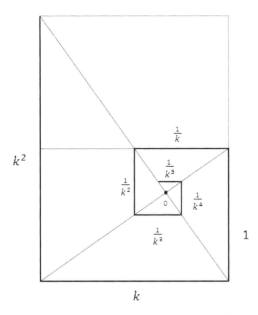

Fig. 15.8.   Recovering the vertex points on a log spiral

with the same proportions $a : b$ at a smaller scale and the other being the gnomon $ABFE$, i.e., $U = U + G$ as shown in Fig. 15.9b. We see here that this construction creates three similar right triangles $\triangle COB$, $\triangle DOC$, $\triangle EOD$ for which, a, b, c are the hypotenuses of these three similar triangles and they satisfy Eq. 15.1.

Looking again at Fig. 15.9a you will see that the same construction is ready to be carried out at a smaller scale so that $U = U + G + G$. This process can be continued indefinitely so that the unit is tiled by a sequence of whirling gnomons and one final unit, $U = G + G + G + \cdots G + U$ as shown in Fig. 15.9c.

Figure 15.10 shows the vertex points of the log spiral about center point O yielding the double geometric series and recovering the spiral of Fig. 15.4c.

15.7.   Consequences of Dynamic Symmetry

If the gnomon is a square, i.e., $G = S$, what are the proportions of the unit (see Fig. 15.9)? Let the proportions of U be $x : 1$, in which case: $a = x$, $b = 1$, $c = x - 1$.

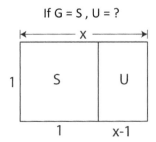

Fig. 15.9.   If the Gnomon is a square, what is the unit?

Then using Eq. 15.1,

$$\frac{x}{1} = \frac{1}{x - 1}$$

or,

$$x - \frac{1}{x} = 1,$$

solving for x,

$$x = \frac{1 + \sqrt{5}}{2} = \phi,$$

which is known as the *golden mean* with many unusual properties, to be explored in the next chapter.

In a similar manner it is easy to show that:

a. When $G = DS$ (double square), the proportions $x : 1$ of the unit satisfies the equation, $x - \frac{1}{x} = 2$ (show this). Solving for x yields $x = 1 + \sqrt{2} = \theta$, known of as the *silver mean*.
b. When $G = U$, then $x = \sqrt{2}$. In other words when a rectangle has proportions: $\sqrt{2} : 1$ (root 2 rectangle), and it is divided in half it yields a pair of root 2 rectangles (show this).
c. When $G = 2U$ then $x = \sqrt{3}$ (show this).
d. When the side of a regular heptagon is 1 unit, the two diagonals are $\rho = 1.801\ldots$ and $\sigma = 2.24\ldots$ and the diagonals subdivide into line segments related to ρ and σ as shown in Fig. 15.10a [Kap13].

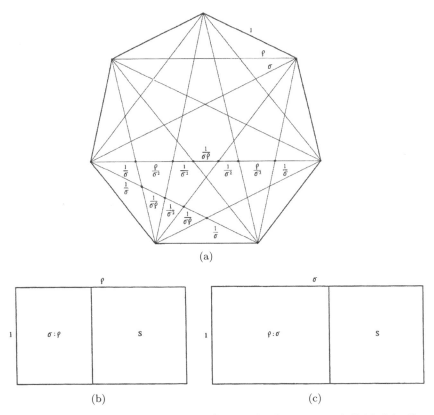

(a)

(b)   (c)

Fig. 15.10. The two diagonals ρ, σ of a regular heptagon subdivided in line segments related to ρ and σ; (b) A $\rho : 1$ rectangle with a square removed yields a $\sigma : \rho$ rectangle; (c) A $\sigma : 1$ rectangle with a square removed yields a $\rho : \sigma$ rectangle

If U has proportions, $\rho : 1$, then G has proportions, $\sigma - 1 : 1$, while if $U = \sigma : 1$, then $G = \rho : 1$.

e. Analogous to the golden mean if a square is removed from a $\rho : 1$ rectangle, it leaves a $\frac{\sigma}{\rho} : 1$ rectangle whereas if a square is removed from a $\sigma : 1$ rectangle it leaves a $\frac{\rho}{\sigma} : 1$ rectangle [Kap2] as shown in Fig. 15.10b and 15.10c. We have discovered that the diagonals of all odd sided regular polygons form algebraic systems with multiplication and division tables [Kap10].

The golden mean lies at the basis of a regular pentagon with the ratio of diagonal to side being Ø with the diagonals dividing themselves in the golden section (Ø:1). The golden mean has many applications to mathematics, art, and architecture [Kap1,2,3], [Liv]. The silver mean lies at the basis of a regular octagon with diagonals dividing themselves in the silver mean. The $\sqrt{3}$ geometry lies at the basis of six and twelve pointed stars [Kap12].

The silver mean governs a geometry based on $\sqrt{2}$ which was used in Roman times to create much of the architecture of the Roman Empire [Kap3], [Wat]. The geometry based on $\sqrt{3}$ was used by Palladio to create some of his houses [Kap12].

Construction 3. Whirling Squares

Since the gnomon of the golden rectangle is a square, quarter circle arcs can be drawn within the sequence of *whirling squares* to form an excellent approximation to a golden logarithmic spiral as shown in Fig. 15.11.

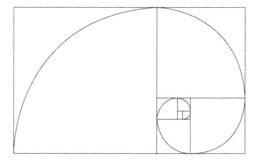

Fig. 15.11.   Whirling squares approximate a golden log spiral

Construction 4. Root 2 Rectangles

Because root 2 (SR2) rectangles have the property that they replicate when divided in half, this results in an interesting geometry [Kap6]. Fig. 15.12 illustrates the dynamic symmetry of a root 2 rectangle showing 4 log spirals. Fig. 15.13 shows a design of two root 2 rectangles juxtaposed at right angles with the regions formed by their guidelines shaded to form an interesting design [Edw].

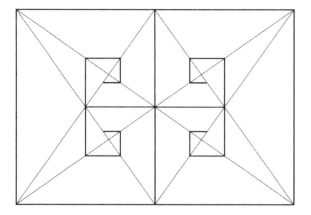

Fig. 15.12.   A spiral design based on a $\sqrt{2}$: 1 rectangle

Fig. 15.13.   Perpendiculalr $\sqrt{2}$: 1 rectangles yield a complex design

Construction 5. The Baravelle Spiral

a. Begin with square *ABCD* and locate the midpoints of *AB*, *BC*, *CD*, and *DA* in Fig. 15.14a.

b. Label the midpoints E, F, G, H respectively.

c. Join the midpoints with line segments to form square *EFGH*, referred to as an ad quadratum square.

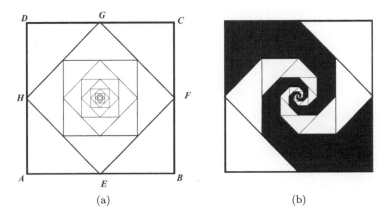

Fig. 15.14.   Steps in the evolution of a Baravelle spiral

d. Similarly, join the midpoints of the sides of *EFGH* to form another square.
e. Repeat the process until the final square is the desired size.
f. The "spiral" shape becomes visible when the triangles are shaded as illustrated in Fig. 15.14b. The resulting figure is a Baravelle spiral.

Remark: If the outer square has an area of 1 unit, the areas of successive squares follow the sequence: 1, 1/2, 1/4, 1/8, ... Since this is a geometric sequence it results in a spiral form.

15.8.   A Musical Connection

The musical scale has a spiral structure which is described in Chap. 21.

15.9.   Conclusion

The Greek quadrivium can be stated as follows:

> To the man who pursues his studies in the proper way, all geometric constructions, all systems of numbers, all duly constituted melodic progressions, the single ordered scheme of all celestial revolutions should disclose themselves ... by the revelation of a single bond of natural interconnection.
>
> — Epinomis

We have seen that by working with the simplest notion of self-similarity of a right-triangle, principles of astronomy, music, geometry, and number, the elements of the Greek quadrivium reveal themselves. As this was considered the proper way to instruct students from the Greek academy, can it not be used to stimulate the thinking of today's schools?

Appendix 15.A

Table 15.1 illustrates four major properties of logarithms:

i) The logarithm of 1 to any base equals 0, e.g., $\log_2 1 = 0$.
ii) The logarithm of the base equals 1, e.g., $\log_2 2 = 1$.
iii) As two numbers multiply, their logarithms add. For example, while the numbers 2, 3 and 5 from the 2nd column of Table 15.1 are the result of an addition , i.e., $2 + 3 = 5$, the corresponding numbers in the right hand column multiply, e.g., $4 \times 8 = 32$.
iv) As a number is taken to a power, its logarithm is multiplied by that power, e.g., $\log 3^2 = 2 \log 3$.

We have limited ourselves in Table 15.1 to finding logarithms of numbers that are powers of 2. So, how can we compute $\log_2 3$? You may be inclined to use your calculator, but you will be disappointed to find that calculators have no direct way to compute $\log_2 3$. The answer is found in a formula that we introduce without proof,

$$\log_a y = \frac{\log_b y}{\log_b a}. \tag{15.A.1}$$

If we take $b = 10$ then we see that calculators are able to compute logarithms to the base 10, and for $a = 2$ and $b = 10$ we find that,

$$\log_2 y = \frac{\log_{10} y}{\log_{10} 2} = 3.322 \quad \log_{10} y.$$

Therefore,

$$\log_2 3 = 3.322 \quad \log_{10} 3 = 1.585.$$

To summarize the properties of logarithms to any base $k > 0$,

$$\log_k 1 = 0,$$

$$\log_k k = 1,$$

$$\log_k(ab) = \log_k a + \log_k b \text{ and } \log_k \frac{a}{b} = \log_k a - \log_k b,$$

$$\log_k a^b = b \log_k a.$$

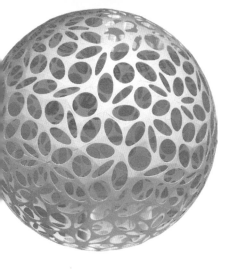

Chapter 16

The Golden Mean

16.1. Introduction

Artists, composers, architects, scientists, and engineers have often created their best works by keeping their eyes open to the natural world. The natural world consists of an elegant dialogue between order and chaos. Careful study of a cloud formation or a running stream shows that what at first appear to be random fluctuations in the observed patterns are actually subtle forms of order. Mathematics is the best tool that humans have created to study the order in things.

Despite the infinite diversity of nature, mathematics and science have attempted to reduce this complexity to a few general principles. In this chapter we introduce you to one enigmatic number, the *golden mean*, which appears and reappears throughout art and science [Kap1], [Kap2], [Liv], [Sta]. We show that this number is actually a member of a family of numbers known as silver means which have applications to architecture and design as well as mathematics and science and this will be discussed in Chap. 20.

16.2. The Golden Mean and the Golden Section

In the last chapter we encountered a number $\phi = \frac{1+\sqrt{5}}{2}$, known as the *golden mean*. The story of the golden mean begins with an undifferentiated line segment.

The line segment is divided into two parts such that the ratio of the whole to the largest segment equals the ratio of the larger to the smaller as shown in Fig. 16.1.

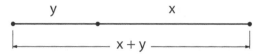

Fig. 16.1.   Golden section as the ratio of the whole line segment to the larger part equaling the ratio of the larger part to the smaller

From this simple statement, all follows about this extraordinary number. This is consistent with Gordon Spencer Brown's notion that once you make a mark on an undifferentiated medium this determines the universe of discourse that unfolds [Spe-B].

First of all, the above statement can be written in terms of three equations where x is the larger segment and y the smaller (make sure that you understand where these equations come from):

$$\frac{x+y}{x} = \frac{x}{y} \tag{16.1}$$

$$z - \frac{1}{z} = 1 \quad \text{where } z = x/y \tag{16.2}$$

$$1 + z = z^2. \tag{16.3}$$

The positive solution to these equations is $z = \phi = \frac{1+\sqrt{5}}{2}$, the *golden mean*, and so the original undifferentiated line segment is divided into a ratio of $1 : \phi$ referred to as the *golden section*. Some time ago I discovered a wonderful construction of the golden section in the notebooks of the artist Paul Klee. It is displayed in Fig. 16.2 where $AC : AB = 1 : 2$. It can be constructed as follows:

Construction 1: Golden section to divide a given line segment into two line segments the ratio of those whose lengths are the golden mean, i.e., in the golden section, proceed as follows (see Fig. 16.2).

a. Start with a line segment AB.
b. Draw $AC = 1/2$ AB at right angles to AB.
c. Circular arc of radius CA intersects CB at F.

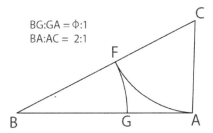

BG:GA = Φ:1
BA:AC = 2:1

Fig. 16.2.   Paul Klee's construction of the golden section

d. Circular arc of radius *BF* intersects *AB* at *G* breaking *AB* into a
golden section.

Equation 16.1 expresses the fact that the golden mean creates a proportion in which the whole is similar to its parts. In Chaps. 13 and
14 this will be seen to be the principal characteristic of a fractal.

Some would say that the golden mean is "number one" in terms
of its importance. This statement can be justified by taking Eq. 16.2
and rewriting it as,

$$\phi = z = 1 + \frac{1}{z} = 1 + \cfrac{1}{1 + \cfrac{1}{z}} = 1 + \cfrac{1}{1 + \cfrac{1}{1 + \cfrac{1}{z}}}$$

$$= \cdots = 1 + \cfrac{1}{1 + \cfrac{1}{1 + \cfrac{1}{1 + \cfrac{1}{1 + \cdots}}}} \tag{16.4}$$

Cutting off these continued fractions at different levels yields the
set of approximations,

$$1, 2, 3/2 = 1.5, 5/3 = 1.667, 8/5 = 1.6, 13/8 = 1.625, \ldots$$

These are the ratios of successive terms from the *Fibonacci* F-
sequence,

$$1 \ 1 \ 2 \ 3 \ 5 \ 8 \ 13 \ 21, \ldots \tag{16.5}$$

These ratios approximate better and better, in terms of decimal values, the golden mean $\phi = \frac{1+\sqrt{5}}{2}$ which is the irrational number (non-repeating decimal) 1.61803... We say that ϕ is the *limit* of the sequence of ratios of successive terms of Sequence (16.5). This is illustrated graphically in Fig. 16.3. Here we begin with a 1:1 square, the augment it to a 2:1 rectangle. Further augmenting by 2:3, 3:5, 8:5, etc., the rectangle more closely approximates a ϕ:1 rectangle.

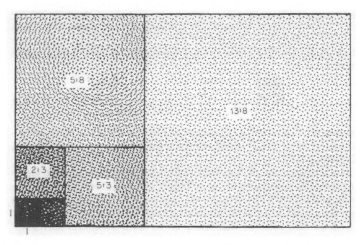

Fig. 16.3.   A 1:1 square can be expanded in an F-sequence (see Sequence 16.50) to a rectangle that asymptotically approximates a golden rectangle

Problem 1: How far out in the Fibonacci Sequence 16.5 do you have to go to get agreement with the golden mean to five decimal places?

A Fibonacci sequence is a sequence of integers that satisfies the recursion formula,

$$a_n = a_{n-1} + a_{n-2} \qquad (16.6)$$

which states that each integer of the sequence is the sum of the previous two numbers. This sequence is very much connected with the growth of plants and other natural phenomena [Kap1], [Kap2]. The additive nature of this sequence led the architect LeCorbusier to use it as the basis of a system of proportions that he called the Modulor [Kap1] which will be discussed in Chap. 18.

In Fig. 16.4, counting the 1's at each level results in the F-sequence. But notice that this pattern of 1's is repeated over and over at different levels. For this reason the F-sequence is said to be self-similar.

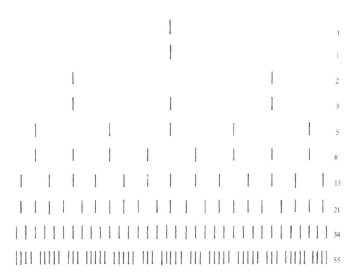

Fig. 16.4.   Diagram illustrating the self-symmetry of the F-sequence

Problem 2: Create a pattern from the Fibonacci numbers $\{1, 2, 3, 5, 8, \ldots\}$. Your pattern can be dots, lines, or anything else of your choosing that gives a geometrical rendering of the Fibonacci sequence.

16.3. The Golden Spiral

In Sec. 15.5 when $k = \phi$, the *golden mean*, this results in a spiral just like the one in Fig. 15.4. The vertex points on the *golden logarithmic spiral* again form a double geometric ϕ-Sequence:

$$\cdots \frac{1}{\phi^3} \quad \frac{1}{\phi^2} \quad \frac{1}{\phi} \quad 1 \quad \phi \quad \phi^2 \quad \phi^3 \quad \phi^4 \cdots \tag{16.7}$$

This geometric sequence has another wonderful property. It is also a Fibonacci sequence, i.e., each number in the sequence is the sum of the two previous numbers. For example

$$\frac{1}{\phi^2} + \frac{1}{\phi} = 1, 1 + \phi = \phi^2, \quad \text{etc.}$$

Problem 3: Show by algebra that,

a. $\left(\dfrac{1 + \sqrt{5}}{2}\right)^2 = 1 + \left(\dfrac{1 + \sqrt{5}}{2}\right),$

b. $\dfrac{1}{\left(\frac{1+\sqrt{5}}{2}\right)^2} + \dfrac{1}{\left(\frac{1+\sqrt{5}}{2}\right)} = 1.$

16.4. A Golden Rectangle

A golden mean rectangle can be constructed with compass and straightedge as follows (see Fig. 16.5).

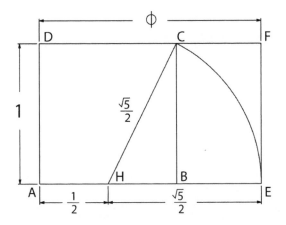

Fig. 16.5.   Construction of a golden rectangle

Construction 2: A Golden Rectangle

a. Begin with a square $ABCD$ of side 1 unit.
b. Bisect the base AB.
c. Extend the base.

d. Place your compass point at the midpoint H of the base and use a compass to sweep out an arc with radius HC until it intersects the extended baseline AB at E.

e. Since $HC = \frac{\sqrt{5}}{2}$, we have that $AE = \phi = \frac{1+\sqrt{5}}{2}$.

Construction 3: The phi sequence

Use Construction 1 to create a golden section, $1 : \phi$. Now that you have lengths 1 and ϕ you can construct all of the terms in Sec. 16.7 by making use of the Fibonacci property. The result is shown in Fig. 16.6. Details of the construction will be given in Sec. 18.2.

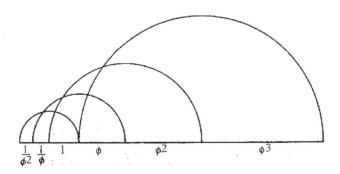

Fig. 16.6.   Construction of the phi sequence

We have also seen in Sec. 15.7 Construction 3, that if you remove a square from a golden rectangle you are left with another golden rectangle at a smaller scale.

Problem 3: Show for Fig. 16.5 that the ratio of sides of the rectangle *BEFC* is $1 : \phi$.

16.5. A Regular Pentagon

You can also use the golden mean to construct a *regular pentagon* since the diagonal of a pentagon has the value ϕ when the length of the side is 1 unit as shown in Fig. 16.7. Therefore, the pentagon can be constructed by the following procedure.

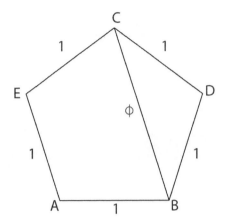

Fig. 16.7. Ratio of the diagonal to the side of a regular pentagon equals the golden mean.

Construction 4: A Regular Pentagon

a. Begin with line segment AB of length 1 unit where A and B are two vertices of the pentagon.

b. Sweep out two arcs from A and B of length ϕ units where 1 and ϕ have been constructed from the golden section as in Construction 1. The two arcs intersect at C, another vertex of the pentagon.

c. Sweep out two arcs at B and C of length 1. Where they intersect is vertex D.

d. Do the same at vertices A and C to form vertex E of the pentagon.

16.6. Golden Triangles

From the pentagon you can construct two isosceles *golden triangles* as shown in Fig. 16.8a. Triangle 1 has base 1 and sides ϕ with base angles equal to 72 deg. and the remaining vertex angle of 36 deg. Triangle 2 has a base of ϕ and sides of 1 unit. Because of this simple geometry, these triangles are self-similar in the following sense. If you bisect the 72 deg. angle of Triangle 1, it results in similar versions of triangles 1 and 2 at a smaller scale (see Fig. 16.8b). This construction can be continued over and over to construct triangles at smaller and smaller scales (see Fig. 16.8c).

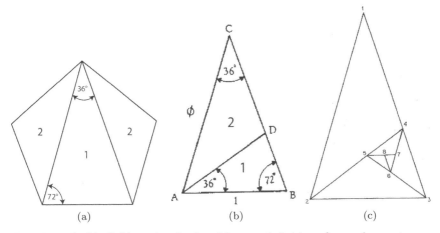

Fig. 16.8. (a, b). Golden triangles 1 and 2 as a subdivision of a regular pentagon; (c) Golden triangles at smaller scale

Construction 5: Construct golden triangles 1 and 2 at three different scales and fit them together into a pleasing design. Two examples of student projects are shown in Fig. 16.9.

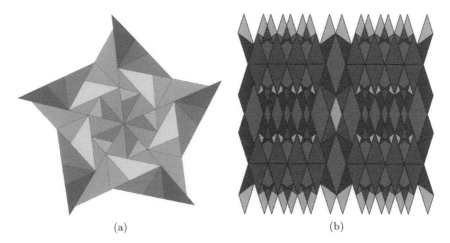

Fig. 16.9. (a, b) Design with golden triangles: from the studio of Jay Kappraff

16.7. Applications of the Golden Section to Art and Architecture

The golden mean is related to human scale since your belly button divides your height approximately in the *golden section*, i.e., the ratio of 1: ϕ as shown in Fig. 16.10 for "Modulor Man," the symbol that the architect, LeCorbusier used to represent his proportional system based on the golden mean known as the Modulor [Kap1]. A construction of the golden section was described in Sec. 16.2 and illustrated in Fig. 16.2.

Fig. 16.10.   Modulor man

Exploration: The golden mean and art

The golden section has appeared in countless works of art as a natural center of tension. It often defines the central point of a painting which is slightly off of the geometric center so as to create a sense of tension or drama in the work of art. In some instances, the artist has intentionally introduced the golden section. In other cases the golden section makes its appearance through some unconscious process on the part of the artist; it may be naturally programmed into our brains.

I invite you to explore great paintings to see if the central point of the painting divides the canvas into the golden section. To help you in this analysis you can use the golden section line chopper in Fig. 16.11 which you may enlarge. You may reprint this figure and enlarge it. The horizontal line on the top of the line chopper is divided in the golden section. From a print of a classic painting take either its length or width and use the line chopper to find its golden section, and see if the central point of interest within the painting divides the canvas in the golden section.

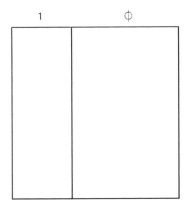

Fig. 16.11.   A golden section chopper

16.8.   A Regular Decagon and Pentagon within a Circle

Construction 6: Construct a regular decagon and pentagon within a circle.

You are now able to inscribe a regular *decagon* inside of a circle as follows:

a. Use Construction 1 to construct golden triangle 1 with base 1 unit and sides ϕ units (see Sec. 16.6). The vertex angle of this triangle is 36 deg., the central angle of the decagon. Therefore, the sides of triangle 1 are radii of the circle and the base is a chord of the circle as shown in Fig. 16.12.

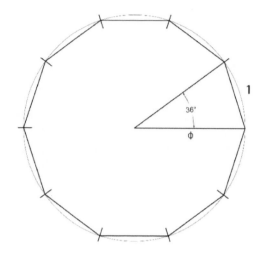

Fig. 16.12.   A decagon created from golden triangles 1

b. Mark off 10 equal chord lengths to construct the sides and vertices
of the decagon inscribed in the circle.

16.9. The Golden Mean and the Tetractys

A structure known as the tetractys goes back to the time of Plato.
The tetractys consists of ten pebbles arranged in rows of 1, 2, 3, 4
as shown in Fig. 16.13a. Adding three additional pebbles creates a
six-pointed star as shown in Fig. 16.13b. It is interesting that the star
is organized by 12 dots surrounding a central dot, just as 12 apos-
tles surrounded Christ, 12 disciples surrounded Mohammad, and 12

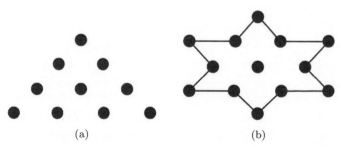

(a)   (b)

Fig. 16.13.   (a) The tetractys; (b) The tetractys creates a star made up of 12
surrounding 1

tribes of Israel surrounded Moses. The tetractys signified the harmony of the universe. After all, the ratios of the first four integers represent the harmonies that comprise the Pythagorean musical scale which became the cornerstone of Western music: the unison, 1:1, the octave or diapason, 2:1, the perfect fifth or diapente, 3:2, the perfect fourth or diatesseron, 4:3, and an diapason above the diapente, 3:1. We devote much of Chap. 21 to describing this musical system and its relationship to architecture. The numbers 1, 2, 3, 4 also signify point, 1, line, 2, plane, 3, and 3-dimensional space, 4, as shown in Fig. 16.14.

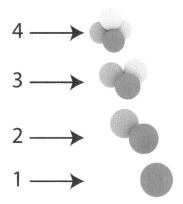

Fig. 16.14.   Spheres representing point, line, plane and 3-D space.

The designer, amateur geometer, and cartoonist, Janusz Kapusta, discovered an extraordinary connection between the tetractys and the golden mean. Place the pebbles of the tetractys at the centers of ten squares as shown in Fig. 16.15a. A pair of lines through the upper left-hand and right-hand corners of the squares define an additional square in Figs. 16.15b and 16.15c. Next, inscribe circles within the squares, tangent to the sides of the squares, including a circle in the top square (see Fig. 16.5d). Now move the inclined lines parallel to themselves until they are tangent to the circles (see 16.15e). Notice, in Fig. 16.15e, that the two inclined lines meet in the top circle at the golden section of the vertical diagonal. The vertex of this triangle intersects within the top circle so that it divides the

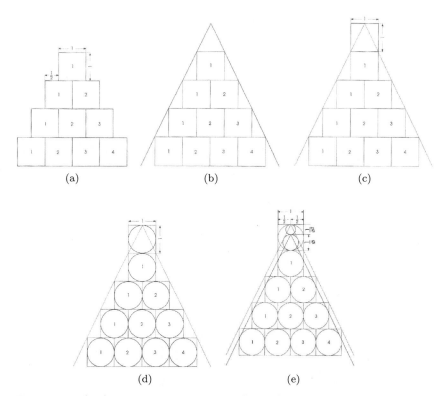

Fig. 16.15.   (a–e) Ten unit squares in the form of a tetractys leads to a pair of circles with diameters: $\frac{1}{\phi}$ and $\frac{1}{\phi^2}$ ([Kap])

vertical diameter of that circle in the golden section $\phi : 1$ represented by a pair of circles with diameters $\frac{1}{\phi}$ and $\frac{1}{\phi^2}$ where $\frac{1}{\phi} + \frac{1}{\phi^2} = 1$.

Next, Fig. 16.16 shows an exploded view of this pair of circles to be the initiating circles of an infinite sequence of "kissing" (tangent) circles with negative integer powers of the golden mean as their diameters show in Fig. 16.17. Finally, an infinite collection of circles in Fig. 16.18 shows that the diameters of the odd inverse powers of the golden mean sum to 1 unit, i.e.

This leads, in Fig. 16.19, to a new construction of the golden section of an edge of the square. I leave it to the reader to carry out this construction and prove its validity.

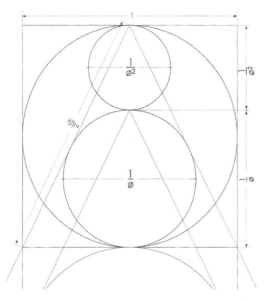

Fig. 16.16.   Detail of the circle in Fig. 16.15e illustrating that $\frac{1}{\phi} + \frac{1}{\phi^2} = 1$ ([Kap])

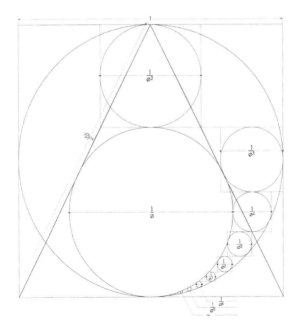

Fig. 16.17.   "Kissing" circles formed circles of diameters with the inverse powers of ϕ ([Kap])

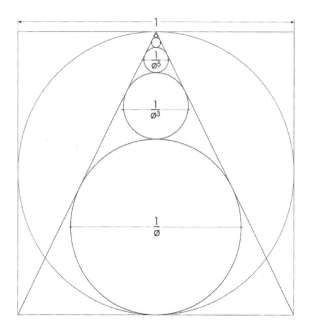

Fig. 16.18.   Sum of the odd inverse powers of ϕ equals 1 ([Kap])

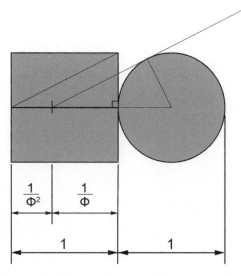

Fig. 16.19.   A new construction of the golden section ([Kap])

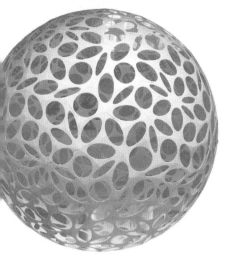

Chapter 17

Wythoff's Nim

My own interest in the fascinating world of the golden mean began as the result of playing Wythoff's Nim [Cox]. The game is played as follows:

Begin with two stacks of tokens (pennies). A proper move is to remove any number of tokens from one stack or an equal number from both stacks. The winner is the person removing the last token.

The winning strategy is based on the following theorem due to S. Beatty:

Theorem 17.1: *If $\frac{1}{x} + \frac{1}{y} = 1$ where x and y are irrational numbers, then the sequences $[x]$, $[2x]$, $[3x]$,... and $[y]$, $[2y]$, $[3y]$,... together include every positive integer taken once ([ ] means "integer part of" for example $[3.4] = 3$).*

Since $\frac{1}{\phi} + \frac{1}{\phi^2} = 1$ for the golden mean, Beatty's theorem shows that $[n\phi]$ and $[n\phi^2]$ exhausts all of the natural numbers with no repetitions, as n takes on the values, $n = 1, 2, 3, \ldots$ Table 17.1 shows results for $n = 1, 2, \ldots, 6$. Do you notice a pattern in these numbers that enables you to continue the table without computation? These Beatty pairs are also the winning combinations for Wythoff's Nim. At any move, a player can reduce the number of counters in each stack to one of the pairs of numbers in Table 17.1 while the other

Table 17.1. Winning combinations of Wythoff's Nim

n	$[n\phi]$	$[n\phi^2]$
1	1	2
2	3	5
3	4	7
4	6	10
5	8	13
6	9	15

player is unable to restore a winning pair. The player who does this at each turn is assured victory.

Notice in the $[n\phi]$ column, the number of pennies in the left hand column progress according to the difference sequence: 21221212...

And in the $[n\phi^2]$ column the number of pennies in the right hand column progresses according to the difference sequence: 32332323...

If 1 is subtracted from each digit in the left-hand column and 2 units are subtracted from the right-hand column you get the sequence, in both cases:

$$10110101 \ldots \tag{17.1}$$

This sequence, which I will refer to as the Golden Sequence, is symbolic of the golden mean and the F-sequence. It appears in many applications of the golden mean [Kap2]. For example, it occurs in the study of plant growth and chaos theory. To see how the F-sequence is embedded in the Golden Sequence (17.1), initiate the sequence by 0. Every time you see a 0, replace it by a 1; whenever you see a 1 replace it by 10. This results in the following Expression (17.2):

$$
\begin{array}{l}
0 \\
1 \\
10 \\
101 \\
10110 \\
10110101 \\
1011010110110
\end{array} \tag{17.2}
$$

Notice that the sequences within Expression (17.2) are approaching the Golden Sequence (17.1). In a sense, the sequences in Expression (17.2) can be thought of as a Fibonacci sequence in that, by juxtaposing two successive sequences from Expression (17.2), results in the next sequence. And, if you count the number of 0's and 1's in each sequence of Expression (17.2), as is done in Table 17.2, you can see the F-sequence evolving in columns 2 and 3?

Table 17.2. Fibonacci nature of Wythoff's Nim

Number	Number of 0's	Number of 1's
1	1	0
2	0	1
3	1	1
4	1	2
5	2	3
6	3	5
7	5	8

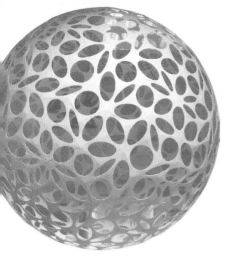

Chapter 18

The Modulor of Le Corbusier

18.1. Introduction

The Modulor is a scale of proportions created by the architect, Le Corbusier. It was modeled after the image of "Modulor Man", a six foot British policeman with his hand over his head and was based on the proportions of the human body. Le Corbusier created this system as a reaction to the increasingly systemization brought about through the use of the metric system. Whereas, the British system of measurement was based on the human body, for example, elbow (cubit), finger (digit), thumb (inch), the metric system was based on the meter which was a forty-millionth part of the meridian of the Earth, way beyond human scale.

Le Corbusier built the Modulor based on the golden mean which is known to be related to human scale and has many additive properties. He described his system in a two-volume set of books, Modulor [LeC], published in 1953. This chapter shows how the Modulor system can be used to tile a rectangle with golden mean proportions. Once a tiling is discovered, the additive properties enable the same set of tiles to tile the same rectangle in many ways, something that would not be possible without the additive properties.

18.2. Golden Mean Exercises

We have seen in Chap. 16 the Fibonacci sequence, or F-sequence,

$$1\ 1\ 2\ 3\ 5\ 8\ 13\dots \tag{18.1}$$

where each term of the sequence is the sum of the preceding two terms. It also has the property that the ratio of successive integers from this sequence approach the golden mean, ϕ, where

$$\phi = \frac{1+\sqrt{5}}{2} = 1.618.$$

Also the golden mean forms a double geometric ϕ-sequence,

$$\dots \frac{1}{\phi^2}, \frac{1}{\phi}, 1, \phi, \phi^2, \phi^3, \dots \tag{18.2}$$

that is also a Fibonacci sequence.

It is because the ϕ-sequence is both geometric and additive that Le Corbusier used it as the basis of a system of architectural proportions called the Modulor (See Chap. 16) [LeC1], [LeC2], [Kap3]. As the result of (18.2) being a Fibonacci sequence,

$$\dots \frac{1}{\phi^2} + \frac{1}{\phi} = 1, \frac{1}{\phi} + 1 = \phi, \quad 1 + \phi = \phi^2, \phi + \phi^2 = \phi^3, \dots \tag{18.3}$$

and in reverse,

$$\dots, \phi - 1 = \frac{1}{\phi}, \phi^2 - \phi = 1, \phi^3 - \phi^2 = \phi, \dots$$

To see how the additive properties work from an algebraic standpoint consider the expression:

$$\phi^2 + 2\phi - 1 - \frac{1}{\phi}.$$

By using the Fibonacci properties of the ϕ-sequence in (18.2) this expression can be transformed as follows:

$$\phi^2 + 2\phi - 1 - \frac{1}{\phi}$$

$$= (\phi^2 + \phi) + (\phi - 1) - \frac{1}{\phi}$$

$$= \phi^3 + \frac{1}{\phi} - \frac{1}{\phi}$$

$$= \phi^3$$

Use the additive properties of the ϕ-sequence given by Expression 18.3 to determine which pairs of the following expressions are equal. Do not use a calculator to solve this problem although you can check your results on a calculator.

Problem 1:

Exercise 1:

a) $2\phi^2 + 1$
b) $\phi^3 - \phi + \phi^2$
c) $\phi + 3 + \frac{1}{\phi}$
d) $2\phi + 3$

Exercise 2:

a) $\phi + \phi^2 - \frac{1}{\phi} + 2$
b) $\phi^2 + 3\phi + 1$
c) $2\phi^2$
d) $2\phi + 2 + \frac{1}{\phi^2}$

Problem 2: One of the two diagrams in Fig. 18.1a or Fig. 18.1b, is not consistent in its dimensions. Which is it and why? (Note: the diagrams are not drawn to scale.)

Problem 3: Find a tiling of the rectangle in Fig. 18.2 into three tiles with two of them golden mean rectangles (sides in proportion $\phi : 1$) and one square. The rectangle is not drawn to scale.
The ϕ-sequence can be constructed using compass and straightedge as shown in Sec. 16.2 with details as follow:

a) Divide a line segment into the golden section to obtain a pair of lengths in the proportion of 1 and ϕ units (See Sec. 16.2).
b) Mark the two units on a straight line.

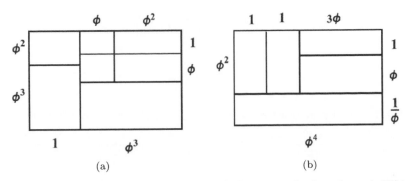

Fig. 18.1.   (a, b) Two tilings of a rectangle by rectangles based on ϕ. Which subdivisions is correct?

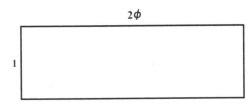

Fig. 18.2.   Tile this rectangle into 3 tiles with two of them golden mean rectangles (proportion ϕ:1) and 1 square

c) With compass and straightedge you can add lengths 1 and ϕ to get length ϕ^2.

d) Lengths ϕ and ϕ^2 can then be added to form ϕ^3.

e) Continuing in this way, all the lengths of the phi sequence can be constructed, leading to Fig. 18.3. Details are left to the reader.

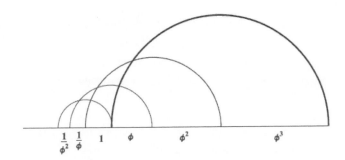

Fig. 18.3.   Construction of the phi-series

18.3. The Modulor of Le Corbusier

1. The Modulor of Le Corbusier is a double sequence of scales referred to as the Red and Blue series. These scales are illustrated in Fig. 18.4a, not drawn to scale.

Blue | $\quad\dfrac{2}{\phi}\quad 2\quad 2\phi\quad 2\phi^2\quad 2\phi^3$

Red | $\qquad 1\quad \phi\quad \phi^2\quad \phi^3\quad \phi^4$

(a)

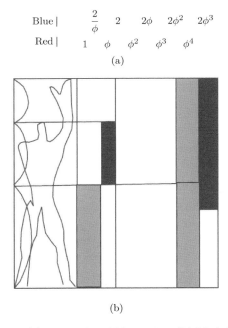

(b)

Fig. 18.4.   (a) The red and blue series, (b) Modulor man

2. Le Corbusier conceived of the Modulor as embodying human scale so that 1 unit from the Red series was taken to be a six foot British policeman while $\frac{2}{\phi}$ units from the Blue series was the measure of the policeman with his hand raised over his head as in Fig. 18.4b.

3. Each scale is a ϕ-sequence. However, each length of the Blue sequence is twice the length of a corresponding length of the Red sequence. Notice how the lengths of each scale intersperse the other. In fact, each Red length bisects the two lengths of the Blue sequence that brace it, i.e., it is the arithmetic mean of the two elements of the blue sequence. However, each length of the Blue sequence divides the line segment of the two bracing Red lengths in the golden section ($\phi : 1$). Show this by using the algebraic

properties of the ϕ-sequence given in Eq. 18.3 to prove the following identity:

$$\frac{2\phi - \phi^2}{\phi^3 - 2\phi} = \frac{1}{\phi}.$$

Another way of saying this is that each element of the blue sequence is the harmonic mean of the two elements of the red sequence that brace it from below, e.g.,

$$2\phi = \frac{2(\phi^2)(\phi^3)}{\phi^2 + \phi^3} \quad \text{(prove this)}$$

where we have used the formula for harmonic mean, b, of a and c, $b = 2ac/(a + c)$.

The lengths from the Red and Blue series were used by Le Corbusier to form the lengths and widths of a modular set of rectangular tiles shown in Fig. 18.5. In this figure there are three types of rectangles. Type 1 is made up of lengths and widths from the Red series, type 2 uses lengths and widths from the Blue sequence, while type 3 uses a length from the Red and a width

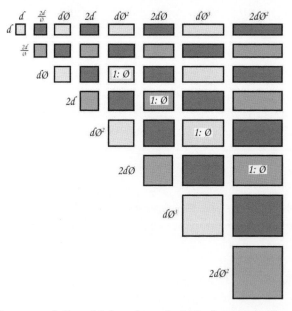

Fig. 18.5.   Sequence of tiles with lengths and widths from the red and blue series

from the Blue or vice versa. These tiles can be used to partition any rectangular space such as a room or the façade of a building.

4. Figure 18.6 has been taken from Le Corbusier's book, Modulor. Each rectangle has been subdivided into Red and Blue sequence rectangles coded according to the rectangles on the upper right. Use the Fibonacci properties of the ϕ-sequence to show that the sum of the lengths across the top edges of the rectangles in Fig. 18.6 agree with the sum of the lengths across the bottom edges. Also check the left and right edges for agreement.

Example: For the rectangle on the upper left-hand corner of Fig. 18.6:

$$L_6 + L_3 + L_1 = L_{11} + L_4$$

$$1 + 2\phi + 2\phi^2 = 2\phi^2 + \phi^3$$

$$\phi^2 + \phi + 2\phi^2 = 2\phi^2 + \phi^3.$$

Therefore, $\phi^3 + 2\phi^2 = 2\phi^2 + \phi^3$.

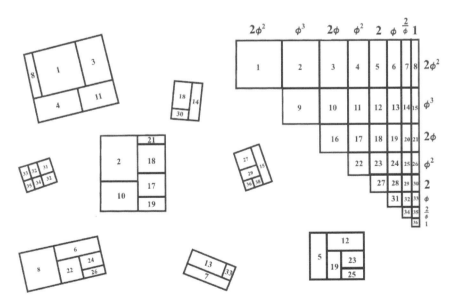

Fig. 18.6. Tiling of rectangles by tiles from the red and blue series

5. Because of the additive properties of the ϕ-sequence, the Modulor is a versatile system of proportions. Figure 18.7 illustrates how the tiles of one particular subdivision of a $\phi^2 \times 2\phi$ rectangle can be rearranged to tile the rectangle in 48 other ways. The original subdivision is not of interest from a design standpoint. However, several of the subdivisions are aesthetically pleasing.

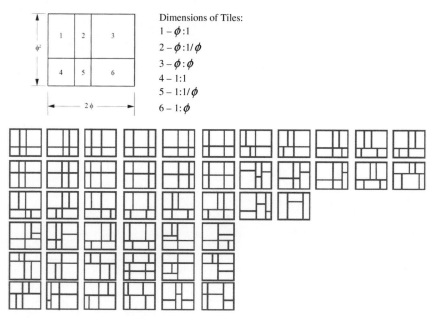

Dimensions of Tiles:
$1 - \phi:1$
$2 - \phi:1/\phi$
$3 - \phi:\phi$
$4 - 1:1$
$5 - 1:1/\phi$
$6 - 1:\phi$

Fig. 18.7. 48 ways of tiling a $\phi^2 \times 2\phi$ rectangle by tiles from the red and blue series

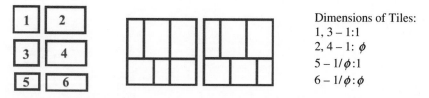

Dimensions of Tiles:
$1, 3 - 1:1$
$2, 4 - 1: \phi$
$5 - 1/\phi:1$
$6 - 1/\phi:\phi$

Fig. 18.8. Six tiles based on ϕ tile two $\phi^2 \times \phi^2$ squares. Can you find 8 additional tilings?

The six tiles shown in Fig. 18.8 with listed dimensions can be combined in many different ways to form a square of dimensions

$\phi^2 \times \phi^2$. Two ways are shown. Cut out several sets of the 6 tiles and combine them in at least 8 other combinations to form squares of dimension, $\phi^2 \times \phi^2$.

6. There are many identities hidden in the Red and Blue sequences. It is sometimes helpful to create a counterpart to this double sequence with the F-sequence and its double as follows (not drawn to scale):

$$2 \quad 4 \quad 6 \quad 10 \quad 16 \ldots \atop 1 \ 2 \ 3 \ 5 \ \ 8 \quad 13 \ldots \tag{18.4a}$$

corresponding to the Red and Blue sequence:

$$\begin{array}{llllll} \text{Blue} & & \frac{2}{\phi} & 2 & 2\phi & 2\phi^2 \ldots \\ \text{Red} & \ldots & 1 & \phi & \phi^2 & \phi^3 & \phi^4 \ldots \end{array} \tag{18.4b}$$

where the elements of each sequence intersperse the other.

We are now able to find number relationships within the integer sequences which lead to the Hidden Identities within the Red and Blue sequences. An example is shown as follows:

a) $1 + 2 + 3 = 6$ corresponds to the Integer Pattern:

$$\begin{array}{cccc} & \text{x} & \text{x} & \circledX \\ \circledX & \circledX & \circledX & \text{x} \end{array} \tag{18.5a}$$

Note that the circles with the x's correspond to the integers from the double sequence (18.4a), e.g., the integers 1,2,3, and 6.

Applying the same pattern to the Red and Blue sequences results in the following Hidden Identity:

$$1 + \phi + \phi^2 = 2\phi^2 \quad \text{(prove this identity)}. \tag{18.5b}$$

b) $3 + 4 + 6 = 13$ corresponds to the Integer Pattern:

$$\begin{array}{ccccc} \circledX & \circledX & \text{x} & \text{x} & \\ \circledX & \text{x} & \text{x} & \circledX & \text{x} \end{array} \tag{18.6a}$$

The Integer Pattern (18.6a), expressed in terms of the Red and Blue Sequences (18.4b), results in the Hidden Identity:

$$\phi + 2 + 2\phi = \phi^4 \text{ (prove this identity).} \qquad (18.6b)$$

c) The same Integer Pattern (18.6a), now using a different starting position for the Red and Blue sequences, results in Hidden Identity:

$$\phi^2 + 2\phi + 2\phi^2 = \phi^5 \text{ (prove this identity).} \qquad (18.7)$$

Problem 4: Choose a rectangle in Fig. 18.6, and show that the lengths of the left and right edges agree and that the lengths of the top and bottom edges agree.

Problem 5: For several integer identities, using the double sequence (18.4a), create the Integer Pattern and the corresponding Red and Blue identity. Using the additive properties of (18.3), prove that the identity is correct.

18.4. A Workshop on the Modulor

Materials needed:

a) Enlargements of the rectangles in Fig. 18.5,
b) Scissors,
c) Glue stick,
d) Magic markers.

Follow these steps:

1. Make several copies of each enlarged rectangle from Fig. 18.5.
2. Draw three squares of dimensions $2\phi^3 \times 2\phi^3$ at the scale of our rectangles.
3. Tile a square of dimensions $2\phi^3 \times 2\phi^3$ using tiles from the Red and Blue sequences.
4. For each tiling, use the ϕ-sequence to show that the sum of the lengths along the top equals the sum of the lengths along the bottom of the tiling and the same for the lengths along the left and right sides.

5. Rearrange the same tiles to get other tilings of the $2\phi^3 \times 2\phi^3$ square.

6. When you get an interesting tiling, use glue stick to make it permanent and color it with magic markers.

Project: Create three different subdivisions of a $2\phi^3 \times 2\phi^3$ square and tile the square in three different ways using each subdivision. Finally tile a 5 inch by 5 inch square to within a quarter inch tolerance using rectangles from the Red and Blue sequences. One example of this project is shown in Fig. 18.9.

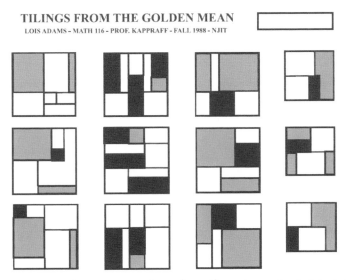

Fig. 18.9. Three different tilings of a $2\phi^3 \times 2\phi^3$ square, each tiled in 3 different ways along with 3 tilings of a 5in. × 5in square to within 1/4 in. tolerance by Lois Adams

Figure 18.10 shows the façade of Unite' D'habilitation in Paris in which Le Corbusier made use of the Modulor.

Fig. 18.10.   Unite' Habilitation incorporating the Modulor

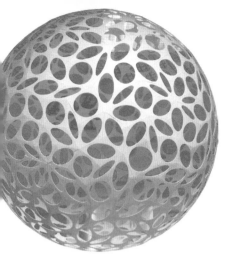

Chapter 19

Non-periodic Tilings of the Plane

19.1. Introduction

A *periodic tiling* is one in which the entire configuration can be translated, without rotation, to a new position which reproduces the original tiling. We say that the tiling is invariant under translation. It was thought until the 1960s that any set of polygons that tile the plane non-periodically could also tile the plane periodically by rearranging the tiles. For example, the spiral tiling by Heinz Voderberg in Fig. 19.1c has been tiled by combining a pair of enneagons (nine sided figures), Fig 19.1a and 19.1b, to form an octagon that fills the plane periodically but also a spiral that tiles non-periodically. It was therefore a great interest that met Robert Berger's discovery in 1964 that there exists a non-periodic tiling of the plane for which there is no periodic tiling. However, to carry out this tiling Berger needed 20,000 tiles.

19.2. Kites and Darts

This enables us to better appreciate Roger Penrose's discovery in the 1970's of two tiles, a kite and a dart. The kite and dart are constructed from a pentagon with side 1 and diagonal ϕ as shown in Fig. 19.2. The kite is constructed from two golden triangle 1's while the dart is constructed from two golden triangles 2 as shown

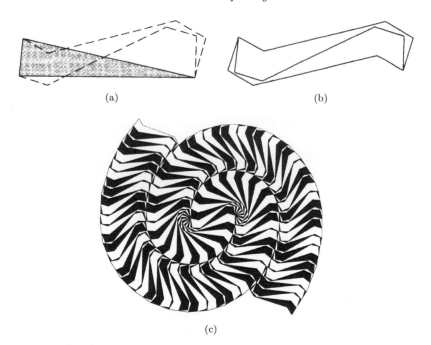

Fig. 19.1. (a, b) Two enneagons (9-sided polygons) (c) A spiral by Heinz Voder-berg by enneagons

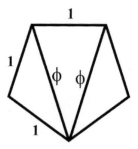

Fig. 19.2. A regular pentagon tiled by 3 golden triangles

in Fig. 19.3 (see Sec. 16.6). You will notice that both kite and dart exhibit red and blue curved lines related to the golden mean. They are guaranteed to tile the plane non-periodically if certain rules are followed stating how these tiles can be combined. (Note that each tile, separately or together, tile the plane periodically if no restrictions are imposed). The blue line of a kite must be placed so that blue line

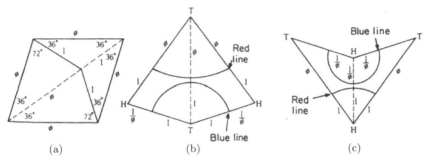

Fig. 19.3. (a, b, c) Kites and darts with blue and red lines (by M. Gardner [Gar4])

meets blue line of another kite or dart, and, similarly, the red line meets red line. In this manner the final tiling has red and blue lines winding through the tiling.

In order to become familiar with these tilings it is helpful to first construct the three combinations of kites and darts shown in Fig. 19.4 referred to as Ace, Short Bowtie, and Long Bowtie. Some starting positions are the Deuce, Queen, and Jack shown in Fig. 19.5. These combine to form the infinite Sun pattern in Fig. 19.6a and the infinite Star pattern in Fig. 19.6b. Using the "Batman" starting position you can create the grand Cartwheel pattern in Fig. 19.6c.

All of these tilings have approximate pentagonal symmetry. This is nicely illustrated in the Sun, Star, and Cartwheel tilings. The tilings are approximately regular in the sense that they seem to

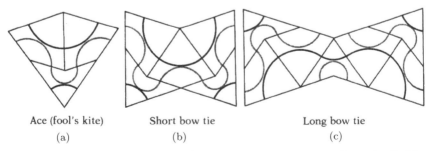

| Ace (fool's kite) | Short bow tie | Long bow tie |
| (a) | (b) | (c) |

Fig. 19.4. (a, b, c) Ace (fools kite), Short bow tie, and Long bow tie tiled by kites and darts with red and blue lines forming a continuous curve (by M. Gardner [Gar4])

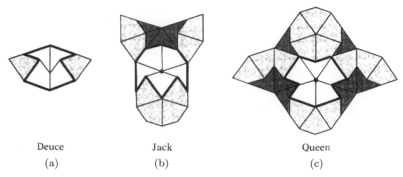

Deuce   Jack   Queen

(a)   (b)   (c)

Fig. 19.5.   (a, b, c) Deuce, Jack and Queen formed by kites and darts (by M. Gardner [Gar5])

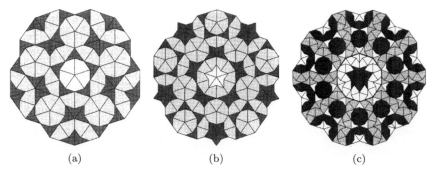

(a)   (b)   (c)

Fig. 19.6.   (a, b, c) Sun pattern, Infinite star pattern, grand Cartwheel pattern (by M. Gardner [Gar5])

always be striving to reproduce themselves but never quite succeeding. Wherever we look, we see a configuration that looks familiar in the sense that we have seen something just like it in one or another position within the tiling. To make this statement more precise we state a remarkable theorem due to Conway. In simple terms, the theorem can be described as follows.

Theorem 19.1: *Let's say you are residing in a finite circular region of diameter d of a Penrose tiling which I refer to as "your universe." Let's call this "your town." If you are transported to another position in your universe, how far must you wander to find an exact replica of your town? Conway proved that in a circle of any position of radius 2d there will always be an exact replica of your town.*

Construction 1

Construct a pattern of at least fifty kites and darts. The following discussion by Martin Gardner in Scientific American [Gar4] may be helpful.

"To appreciate the full beauty and mystery of Penrose tilings one should make at least 100 kites and 60 darts. The pieces need be colored on one side only. The areas of the two shapes are in the golden ratio. This also applies to the number of pieces you need of each type. You will need 1.618. . . as many kites as darts. In an infinite tiling this proportion is exact.

A good plan is to draw as many kites as you can on one sheet, with a ratio of about five kites to three darts using a thin line for the curves. The sheet can be photocopied many times. The curves can then be colored with, say, red and blue felt tipped pens. Conway has found that it speeds constructions and keeps patterns more stable if you make many copies of the ace and bowties. As you expand a pattern, you can continually replace darts and kites with aces and bowties. Actually an infinity of arbitrary large pair of shapes made up of kites and darts will serve for tiling any infinite pattern.

A Penrose pattern is made by starting with darts and kites around one vertex and then expanding radially. Each time you add a piece to an edge you must choose between a kite or a dart. Sometimes the choice is forced. Sometimes it is not. Sometime either piece fits, but later you may meet up with a contradiction, (a spot where no piece can be comfortably added, and be forced to go back and make the other choice). It is a good plan to go around a boundary, placing all of the forced pieces first. They cannot lead to a contradiction. You can then experiment with unforced pieces. It is always possible to continue forever. The more you play with the pieces, the more you will become aware of "forcing rules" that increase efficiency. For example, a dart forces two kites in its concavity, creating the ubiquitous ace."

That Penrose tilings are pentagonal by nature is shown in Fig. 19.7 by the illustration of John Stachura. What Stachura did was to keep the blue and red circular arcs but cleverly eliminate the boundaries of the kites and darts.

Fig. 19.7.   Pentagonal star tiled by kites and darts by John Statura

19.3. Golden Diamonds

As an alternative to kites and darts, non-periodic tilings can be created from a pair of golden diamonds using matching rules shown in Fig. 19.8. The single and double arrows must match in a proper tiling. The golden diamonds are formed from a pair of pentagon triangle 1's and 2's (see Sec. 16.6). One such tiling was created by Macarena Maldonado and is shown in Fig. 19.9.

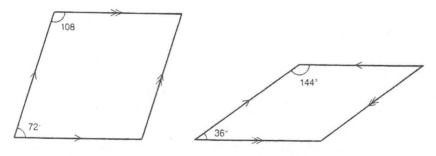

Fig. 19.8.   Golden rhombii with matching rules formed by arrows

GIRAH TILING

Fig. 19.9.   Design with golden rhombii by Macarena Maldonado

19.4. Girih Tilings

A third option was experimented with by Robert Dewar [Dew] who used a family of non-periodic tilings, that may have been antici-pated by Islamic designers of the 12th century A.D., called Girih tiles. Dewar shows that two tiles, a bobbin and a bowtie, can be tiled by kites and darts (see Fig. 19.10) and that these tiles rapidly lead to wonderful non-periodic designs such as the one in Fig. 19.11 by Kwasi Amonkona. The bobbin and bowtie are surrounded by six equal sides,

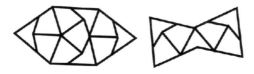

Fig. 19.10.   A bowtie and bobbin tiled by kites and darts

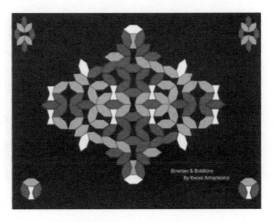

Fig. 19.11.   A tiling by bowties and bobbins by Kwasi Amonkona

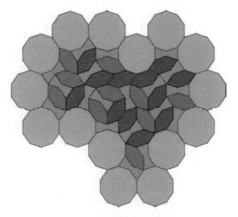

Fig. 19.12.   A tiling with bowties, bobbins and decagons by Nicholas Degio-
vacchino

and their angles are the internal and external angles of decagons (see
Sec. 16.8) in which 3 bobbins and 1 bowtie form a decagon as shown
in the tiling by Nicholas Digiovacchino in Fig. 19.12.

Construction

1. Reproduce and then cut out a number of bobbins and bowties
   (1.618 ... more bobbins than bowties) found in Fig. 19.20. By juxtaposing the bobbins and bowties, try your hand at constructing
   Girih tilings. You will notice that there are designs on the bobbins

and bowties known as grill-work or straps. Information on how to create this grill-work is given in Section 19.5. In your tiling with bowties and bobbins, try to arrange the tiles so that the grill-work is continuous. You can then include the bobbins and bowties in the design or eliminate them leaving continuous strapwork designs. Two such designs are found in Fig.19.21.

2. As described above, 3 bobbins and 1 bowtie can be used to create a decagon. A successful class of Girih tilings is suggested by Dewar in which a ring of decagons is formed in which decagons are attached one to the other by surrounding the decagons by three non-adjacent bowties. Whenever such a ring can be found, Dewar conjectures that the interior can always be tiled by bobbins and bowties as illustrated in Fig. 19.12 and Fig. 19.13. Test this conjecture by enlarging and cutting out the bobbins and bowties at the end of this chapter.

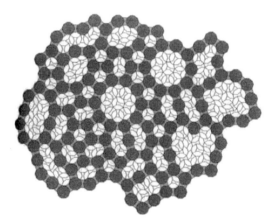

Fig. 19.13. Circuits of decagons enclosing tilings by bowties and bobbins (by Robert Dewar [Dew])

3. The super Girih tiles, shown in Fig. 19.14 can facilitate the Girih tilings.

19.5. The Five Girih Tiles

This section introduces another approach to Girih tiles. You can find more details of this by going to Google.com and placing in the

Fig. 19.14.   Super Girih tiles (by Robert Dewar [Dew])

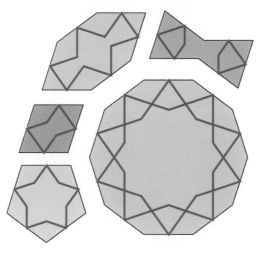

Fig. 19.15.   Five color coded Girih tiles: pentagon, decagon, bowtie, bobbin and rhombus showing strapwork

command line: Girih Tiling Patterns in Google Sketchup. The Girih tilings go as far back as the 13th century. It was originally assumed that Islamic patterns found on the facades of mosques and other sacred structures were created with compass and straightedge. However, a few years ago it was discovered that many of these patterns were made using a set of five tiles. The tiles are shown in Fig. 19.15. They consist of a regular decagon, a regular pentagon, a bobbin (elongated hexagon), a rhombus, and a bowtie, all having the same edge lengths. Their interior angles are given as follows:

Decagon — 144 deg.
Pentagon — 108 deg.

Rhombus — 72 deg., 108 deg.

Bowtie — 72 deg., 72 deg., 216 deg., 72 deg., 72 deg., 216 deg.

Bobbin (elongated hexagon) — 72 deg., 144 deg., 144 deg., 72 deg., 144 deg., 144 deg.

In addition, each of these tiles have decorations referred to as strapwork.

A single guideline is shown emanating from the midpoint of an edge of a Girih tile at 54 deg. to the edge in Fig. 19.16a. In Fig. 19.16b, two pairs of parallel guidelines meet each tile at its midpoint at 54 deg. for a rhombus. The excess guidelines are then

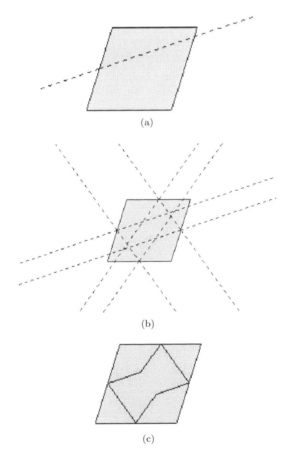

(a)

(b)

(c)

Fig. 19.16.   (a, b, c) Creation of the strapwork

eliminated to yield the strap pattern for the rhombus as shown in Fig. 19.16c. Carrying out this construction can be rather messy. A computer program called "Google Sketch Up" has been created to make this construction more manageable. The web address is:

For PC users http://www.3dvinci.net/SketchUp_Intro_PC.pdf

For Mac users: http://www.3dvinci.net/SketchUp_intro_MAC.pdf

The Sketch Up program permits easy assembly of the tiles into patterns such as the one in Fig. 19.17. Six additional Girih compositions are shown in Fig. 19.18. Try your hand at replicating these patterns.

The Seljuk Mama Hatun Mausoleum in Tercan Turkey exhibiting Girih tiles from the research of Peter Lu and Paul Steinhardt is shown in Fig. 19.19. Notice the Girih tiles that have been superimposed over the straps. This work was published in a paper in the New York Times by John Noble Wilford [Wil]. Emil Makovicky from the University of Copenhagen, in 1992, discovered similar findings independently on a mosaic pattern from a tomb in Iran.

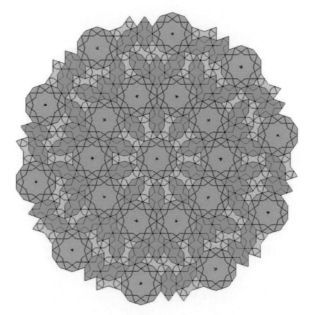

Fig. 19.17.   Assembly of a pattern by the five color-coded Girih tiles using Google Sketchup

Fig. 19.18. Six patterns created by Google Sketchup

Construction:

1. Since non-periodic tilings of the plane can be carried out with bobbins and bowties alone, Fig. 19.20 illustrates a collection of bobbins and bowties with their internal strapwork. Reproduce Fig. 19.20, cut out a collection of bobbins and bowties, and use

Fig. 19.19.   The Seljuk Mama Hatun Mausoleum exhibiting the strapwork over-
laying a pattern of Girih tiles

them to create tilings such as the one in Fig. 19.21a by Kevin
Miranda. Notice how the strapwork joins together continuously.

2. Next remove the outlines of the bobbins and bowties, as was
   done in Fig. 19.21b, and notice how the strapwork flows continuously through the tiling similar to Fig. 19.19 on the wall of
   the Seljuk Mama Hatun Mausoleum discovered by Peter Lu and
   Paul Steinhardt.

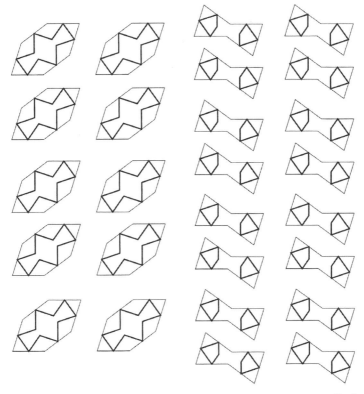

Fig. 19.20.   Bobbin and Bowtie templates to cut out and assemble in Girih tilings

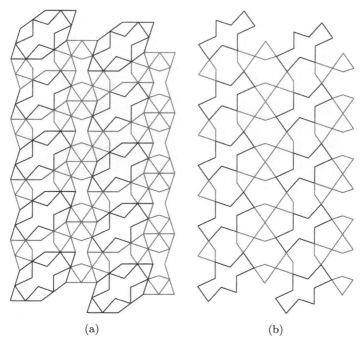

(a)   (b)

Fig. 19.21.   (a) a Girih tilings with bobbins, bowties and internal straps; (b) Tilings with the bobbins and bowties removed, by Kevin Miranda

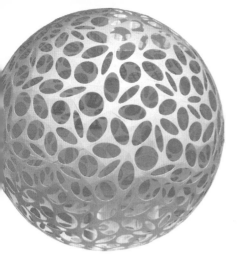

Chapter 20

The Silver Mean

20.1. Introduction

We are looking for a set of tiles with additive properties. Such systems are as rare as hens teeth. The golden mean was shown to be such a tiling in Chap. 18 on the Modulor of Le Corbusier. A system based on a number known as the silver mean is a second and will be discussed in this chapter [Kap6].

In Chap. 15 we encountered a number, $\theta = 1 + \sqrt{2}$ known as the silver mean that satisfies the equation,

$$z^2 - 2z - 1 = 0$$

which can be rewritten as,

$$z - \frac{1}{z} = 2. \qquad (20.1)$$

It is the second member of a family of numbers known as n-th silver mean that satisfy the equation,

$$z - \frac{1}{z} = n. \qquad (20.2)$$

When $n = 1$ we have the golden mean, while for $n = 2$ we have what is referred to simply as the silver mean.

In Fig. 16.16, I presented a diagram showing that the odd inverse powers of the golden mean sum to 1. In Fig. 20.1, I show that sum of

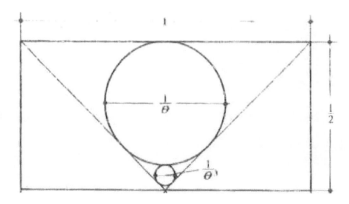

Fig. 20.1.   The sum of the inverse powers of θ equals 1/2 by Janusz Kapusta

the odd inverse powers of the silver mean sums to 1/2. In fact, it can be shown that the sum of the odd inverse powers of the n-th silver mean sums to $1/n$.

Consider the silver mean $\theta = 1 + \sqrt{2}$ which satisfies Eq. (20.1).

By a similar argument to the golden mean made in Sec. 16.2, it can be shown that,

$$\theta = 2 + \cfrac{1}{2 + \cfrac{1}{2 + \cfrac{1}{2 + \cfrac{1}{2 + \cdots}}}} \tag{20.3}$$

Its approximations are the ratio of terms from a sequence known as the Pell sequence:

$$1, 2, 5, 12, 29, 70, \ldots \tag{20.4}$$

satisfying,

$$a_n = 2a_{n-1} + a_{n-2}. \tag{20.5}$$

This sequence was used as the basis of the proportional system used to create much of the architecture of the Roman Empire [Wat]. Theon of Smyrna, a second century A.D. Platonist philosopher and mathematician first presented this sequence in his book The Mathematics Useful for Understanding Plato [The]. The ratio of successive terms from the Pell sequence approaches closer and closer to $\theta = 1 + \sqrt{2}$ in the sense of a limit.

Remark: We saw in Sec. 15.5 that a geometric sequence was characterized by the fact that given three successive terms from the sequence, the square of the middle term equals the product of the first and third, e.g., for the geometric sequence based on 2, $4^2 = 2 \times 8$. In this sense, Sequence 20.4 is an approximate geometric sequence since, given three successive terms, the square of the middle term equals the product of the first and third terms off by 1 unit, e.g., for Sequence 20.4, $5^2 = 2 \times 12 + 1$.

20.2. The Geometry of the Sacred Cut

Like the golden mean, $\phi = \frac{1+\sqrt{5}}{2}$, the silver mean, $\theta = 1 + \sqrt{2}$, leads to a system of proportions that is of interest to designers and architects because of their many additive properties. Let us consider the geometry at the basis of the silver mean. In Fig. 20.2, an arc of a circle is drawn with the compass point at a vertex of a unit square,

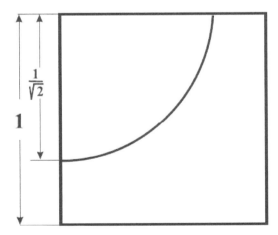

Fig. 20.2. An arc of a circle divides a side in the sacred cut

dividing the side into two parts a and b such that,

$$a = \frac{1}{\sqrt{2}} \quad \text{and} \quad b = 1 - \frac{1}{\sqrt{2}} \quad \text{where} \quad \frac{a}{b} = \frac{\frac{1}{\sqrt{2}}}{1 - \frac{1}{\sqrt{2}}} = \frac{\theta}{1} \quad (20.6)$$

the division of a unit length into two parts in the ratio $\theta : 1$, was referred to by the Danish engineer, Tons Brunes as the *"sacred cut."* Four such arcs initiated from the four vertices of a square, as shown in Fig. 20.3, creates a regular octagon. As shown in Fig. 20.4, using Eq. (20.6), each edge of the square has now been resized into the sacred cut 1 and θ so that the square is divided into three kinds of rectangles with proportion: 1:1, a square (S); $\sqrt{2}$:1, a square root rectangle (SR); and θ:1 that I refer to as a Roman rectangle (RR) shown in Fig. 20.4. Such a subdivision was discovered by Donald and Carol Watts on a tapestry in one of the Garden Homes of Ostia, the port city of the Roman Empire (see Fig. 20.5). Also note that diagonals of the octagon divide themselves in the sacred cut $\theta : 1$.

The additive properties of the Roman system of proportions are reflected in the manner in which these three species of rectangle are related to each other as was done in Sec. 15.6 using Dynamic Symmetry. For example, Fig. 20.6a shows that when S is added or

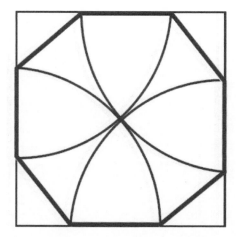

Fig. 20.3.   Four sacred cuts mark the vertices of an octagon

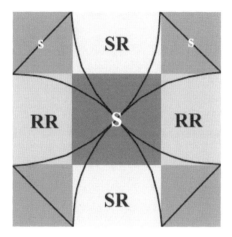

Fig. 20.4. A square divided into three rectangles: a square (S), a square root rectangle (SR) and a Roman rectangle (RR)

Fig. 20.5. A tapestry from Ostia using the tiling in Fig. 20.4

subtracted from an SR it results in a RR. Therefore, in Fig. 20.6b, if a double square (DS) is added to an RR it results in an RR at a larger scale. Also in Fig. 20.6c a pair of SRs results in an SR at a larger scale, whereas, in Fig. 20.6d, cutting an SR in half results in a pair of SRs at a smaller scale. Summarizing these observations,

$$SR \pm S = RR, RR + S + S = RR \quad \text{and} \quad SR + SR = SR. \quad (20.7)$$

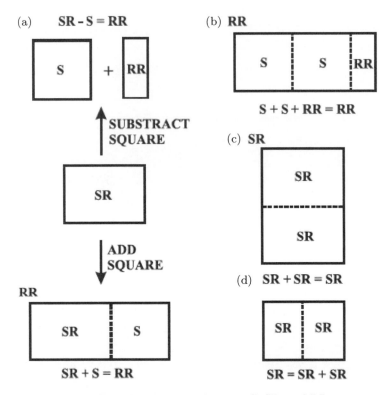

Fig. 20.6.   Relationships between S, SR, and RR

This enables one to begin with an S, and with compass and straightedge create an SR as shown in Fig. 20.7a. Figure 20.7a and Fig. 20.7b can then be used to create S, SR, and RR at three different scales.

By putting two SR's together as in Fig. 20.6d you are able to get a scaled up SR*, and then carry out the scalings in Fig. 20.7 with SR*.

One example of tiling a square with S, SR, and RR at three different scales by Marc Bac is shown in Fig. 20.8.

Construction 1: Create S, SR, and RR at three different scales. Use these rectangles to tile a square or create a pattern.

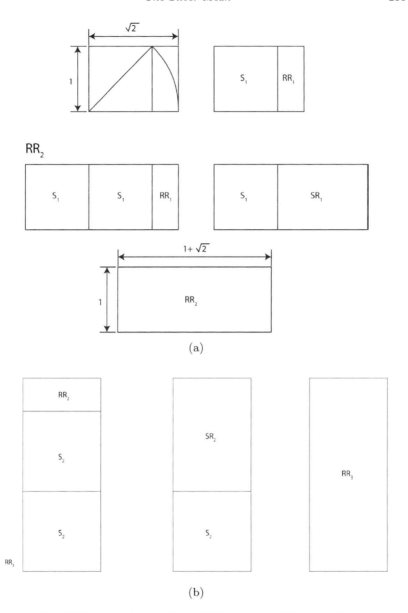

Fig. 20.7. Creating S, SR, and RR at three different scales

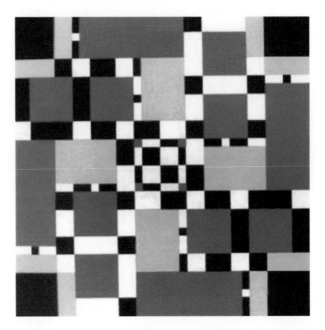

Fig. 20.8.   Tiling of a square by S, SR, and RR by Marc Bak

20.3.   The Arithmetic, Geometric, and Harmonic Means of the Roman System of Proportions

Consider the Pell Sequence (20.4) and a second Pell sequence beginning with 1, 3,

Table   20.1a.   Two   Pell   sequences   approximate $\sqrt{2}$

1	3	7	17	41	99	239	577	...
1	2	5	12	29	70	169	408	...

The ratio of upper and lower integers from this pair of sequences: 1/1, 3/2, 7/5, ... give closer and closer approximations to $\sqrt{2}$ as you go out in the sequence. The ratio 3:2, the musical fifth (see Chap. 21), played an important role in the structure of the Parthenon where it was the proportion of the metope or housing of the statues that surrounded the temple. The ratio 17:12 also figures in the proportions of the Parthenon [Kap12]. The ratio of 577:408 approximates $\sqrt{2}$ to five decimal places. A geometric construction to derive this ratio was recorded in the Sulba Sutra, a geometry book from about 600 BC in Vedic India.

From this double sequence you will notice that,

1. Any integer from the bottom sequence divides the pair of numbers that brace it from above in the *arithmetic mean*, e.g., 5 is the arithmetic mean of 3 and 7.

2. Any integer from the top sequence divides the pair of integers from the bottom sequence that braces it approximately in the *harmonic mean*, e.g., $3 \approx 2(2 \times 5) \div (2 + 5) = 20/7$, where 3 is the approximate harmonic mean between 2 and 5, etc., using Eq. (20.8) for the harmonic mean.

3. Any integer from either sequence is the approximate *geometric mean* of the integer preceding and following it, off by one unit, e.g., $5^2 \approx 2 \times 12 = 24, 12^2 \approx 5 \times 29 = 145$, etc. using the equation below for the geometric mean.

Note: Given the sequence of integers, a, b and c, the arithmetic, harmonic, and geometric means are described by,

$$\text{Arithmetic} : b = \frac{a+c}{2},$$

$$\text{Harmonic} : b = \frac{2ac}{a+c}, \qquad (20.8)$$

$$\text{Geometric} : b = \sqrt{ac}.$$

20.4. The Algebra of the Silver Mean

We described, in Sec. 20.2, the geometry behind the Roman system of proportions based on the Sacred Cut. Now we describe the algebraic system that relates to it. The algebra expresses the many additive properties described by this system. Just as the Red and Blue series of proportions was at the basis of the Modulor of Le Corbusier, the Roman system is based on a sequence of scales. Unlike the Modulor, which requires only two scales, the Red and Blue, the Roman system requires an infinite number of scales for completeness.

Begin with the pair of double-geometric θ-sequences, each being a Pell-sequence:

Table 20.1b. A pair of double geometric θ sequences (Not drawn to scale)

	$\ldots \dfrac{\sqrt{2}}{\theta}$	$\sqrt{2}$	$\theta\sqrt{2}$	$\theta^2\sqrt{2}$	$\theta^3\sqrt{2}$	$\ldots$
$\ldots \dfrac{1}{\theta^2}$	$\dfrac{1}{\theta}$	1	θ	θ^2	θ^3	$\ldots$

Just as with the integer sequences of Table 20.1a, where the ratio of terms approximate $\sqrt{2}$, the ratio of the upper and lower sequences of Table 20.1b exactly equals $\sqrt{2}$. Also, as for the Red and Blue sequence, each number from the bottom sequence fills the gaps created by the upper sequence and provides the arithmetic mean of the two numbers that brace it from above. Also each number in the upper sequence provides the exact sacred cut and harmonic mean of the gap between the two numbers that brace it from below, similar to the integer versions of these sequences exhibited in Table 20.1a.

Recall that the ratio of corresponding terms of the Red and Blue sequence equals 2, i.e., corresponding terms of the Blue sequence are twice the Red. Observing the elements of Table 20.1b, the upper Pell sequence is $\sqrt{2}$ times the lower. Now look at the elements in the first 4 rows of Table 20.2a. Notice that each successive row is the previous row multiplied by $\sqrt{2}$. We can also see this by observing the elements of the integer Table 20.2b. Notice that the upper two rows are double the bottom two rows. It follows that the same holds for Table 20.2a.

Table 20.2a. Four θ-sequences

	$\ldots$	$\dfrac{2\sqrt{2}}{\theta}$	$2\sqrt{2}$	$2\theta\sqrt{2}$	$2\theta^2\sqrt{2}$	$2\theta^3\sqrt{2}$
$\ldots$	$\dfrac{2}{\theta}$	2	2θ	$2\theta^2$	$2\theta^3$	$\ldots$
$\ldots$	$\dfrac{\sqrt{2}}{\theta}$	$\sqrt{2}$	$\theta\sqrt{2}$	$\theta^2\sqrt{2}$	$\theta^3\sqrt{2}$	$\ldots$
$\ldots \dfrac{1}{\theta^2}$	$\dfrac{1}{\theta}$	1	θ	θ^2	θ^3	$\ldots$

Now we turn to the additive properties of Tables 20.2a and Table 20.2b. You can find these additive properties by observing the corresponding integer Table 20.2b. When you find an expression such as, $3 + 2 = 5$, you can translate this to a similar pattern in Table 20.2a, such as $1 + \sqrt{2} = \theta$ or $2 + \sqrt{2} = \theta\sqrt{2}$.

Table 20.2b. Four integer
Pell sequences

2	6	14	34	82	...
2	4	10	24	58	...
1	3	7	17	41	...
1	2	5	12	29	...

There are two fundamental rules, and from these rules you are able to generate all the elements of Table 20.2a with compass and straightedge starting with 1 and $\sqrt{2}$ derived from a square. The rules are stated below along with a third rule pertaining to the property of a Pell Sequence.

Rule 1: The sum of a term and the term above it results in the next term in the sequence following the first term. For example,

$$1 + 1 = 2, 3 + 2 = 5, 5 + 7 = 12, 4 + 3 = 7, \text{ etc.}$$

Applying Rule 1 to Table 20.2a: $1 + \sqrt{2} = \theta$, $\theta + \theta\sqrt{2} = \theta^2$, etc.

Rule 2: The sum of two consecutive terms in any row results in the value above the second term. For example,

$$1 + 2 = 3, 2 + 5 = 7, 7 + 17 = 24, 3 + 7 = 10, \text{ etc.}$$

Applying the rule to Table 20.2a: $1 + \theta = \theta\sqrt{2}$, etc.

Rule 3: This rule mirrors the Pell sequence. For example, using Table 20.2b

$$1 + 2 \times 2 = 5, 2 + 2 \times 5 = 12, 4 + 2 \times 10 = 24, \text{ etc.}$$

Applying this rule to Table 20.2a: $1 + 2\theta = \theta^2$, $\theta + 2\theta^2 = \theta^3$, etc.

Using Rules 1 and 2, all of terms from Table 20.2a can be constructed with compass and straightedge beginning with 1 and $\sqrt{2}$

derived from a square. I will demonstrate this by creating all of the terms from Table 20.2b beginning with $1 + 1$. Using Rules 1 and 2 the terms in the first two rows of Table 20.2b can be constructed as follows:

$$1 + 1 = 2, 2 + 1 = 3, 3 + 2 = 5, 2 + 5 = 7, 7 + 5 = 12, 5 + 12 = 17, \ldots$$

Once the first two rows are complete, the others can be gotten by doubling as seen in Table 20.2b.

In a similar way, all the terms in Table 20.2a can be generated. So one can say that generating the terms of Table 20.2a is as easy as $1 + 1 = 2$.

Problem: Using the fact that $1 + \sqrt{2} = \theta$, use algebra to prove several of the identities exhibited in Rules 1, 2, and 3.

Construction 2: Tile a $2\theta^2 \times 2\theta^2$ square with at least nine rectangles with edges from Sequences in Table 20.2a. Show that these tilings are additive by reorganizing the rectangles to give another tiling using the same set of tiles. This is similar to what you did in tiling a square with rectangles from the Modulor of Le Corbusier in Chap. 18.

For example,

Step 1: Using Table 20.2a, and Rules 1 and 2 to subdivide the top and bottom and the left and right side of the $2\theta^2 \times 2\theta^2$ square.

Top and bottom:

$$2\theta^2 = \theta^2\sqrt{2} + \theta\sqrt{2}$$
$$= \theta + \theta^2 + 1 + \theta$$
$$= 2\theta + \theta^2 + 1$$

Left and right:

$$2\theta^2 = \theta^2 + \theta^2$$
$$= \theta + \theta\sqrt{2} + \theta^2$$

Step 2: Construct with compass and straightedge the six resulting edges using the above construction beginning with 1 and $\sqrt{2}$.

Step 3: Tile the square in Fig. 20.9a with these lengths.

Step 4: Cut out the nine rectangles, numbered 1 through 9, and rearrange them to tile the $2\theta^2 \times 2\theta^2$ square in more interesting ways, one of which is shown in Fig. 20.9b.

This rearrangement can take place because of the many additive properties possessed by the Roman system of proportions. Without additive properties, once you have a tiling, you will never find another one using the same tiles.

Problem: Use algebra to show that the upper equals the lower edge of Fig. 20.9b and the left equals the right edge of Fig. 20.9b.

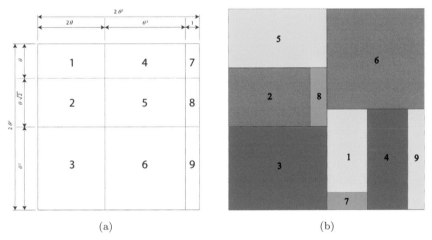

(a) (b)

Fig. 20.9. (a) Tiling of a $2\theta^2 \times 2\theta^2$ square by nine tiles from the Roman system of proportions. (b) Alternative tiling with the nine tiles from Fig. 20.9a

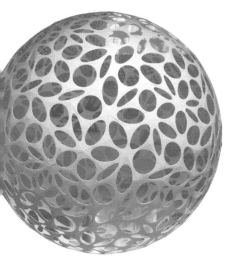

Chapter 21

A Unified Theory of Proportions

21.1. Introduction

The history of proportion in architecture and design has been a search for the key to beauty. Is the beauty of a painting, a vase, or a building due to some qualities intrinsic to its geometry or is it due entirely to the craft of the artist and the eye of the beholder? In this chapter I will examine some of the approaches to proportion that have been used in the past and I will show that they have certain things in common. Throughout this book, theories of proportion have been discussed as entities unto themselves. It is my own discovery that all of these studies are connected and that the nature of the musical scale is at their basis. This chapter is about these connections.

First, we wish to state three canons that most practitioners would agree underlies a good design. All good designs should have:

1. Repetition — some patterns should repeat continuously.
2. Harmony — parts should fit together.
3. Variety — the design should be non-monotonous (not completely predictable).

Many artists and architects would add a fourth requirement that the proportions of a design should relate to human scale in terms of proportions.

A small number of modules should be used over and over rather than fashioning numerous units of disparate size or shape. According to the Renaissance artist and architect Leon Battista Alberti, harmony of proportions should be achieved in such a manner that "nothing could be added, diminished, or altered except for the worse." Any system of proportions should be flexible enough to express the individual creativity of the artist or architect so that the unexpected may be incorporated into the design. Designs from traditional and primitive cultures reflected the desire of people to be connected to their art and their dwellings. An architectural system should also have additive proportions so the floorplan can be subdivided in a way that all dimensions are commensurate with each other; they fit.

We will show how three systems of proportion satisfy these canons: 1) The system of musical proportions from the Renaissance; 2) the Modulor of Le Corbusier, the 20th century French architect, based on the golden mean; and 3) the system of Roman architecture based on the sacred cut and the silver mean.

21.2.   The System of Musical Proportions of Alberti

The Renaissance architect and artist, Leon Battista Alberti, created a system of architecture based on the ratios 2:1 and 3:1 suggested by the Timaeus of Plato in a structure known as the "world soul" shown in Fig. 21.1.

Fig. 21.1.   Plato's World Soul

This structure had its basis in the musical scale. I saw a replica of the World Soul in the Alberti Museum in Florence, Italy. Alberti reasoned that "what is pleasing to the ear should be pleasing to the eye." This pair of ratios can be used to create the ancient musical scale going back to the time of ancient Sumeria and brought to ancient Greece by Pythagoras. The philosophers of the Renaissance used neo-Platonic ideas from ancient Greece to fashion their society and construct their architecture. They based their architecture on a table found in the work of a 2nd century AD mathematician, Nicomachus, who was one of the last mathematicians to record what was known about the musical scale from ancient Greece. The scale of proportions from Nicomachus' table is shown below:

Table 21.1. Nicomachus Table

1	2	4	8	16	32			
	3	6	12	24	48			
		9	18	36	72			
			27	54	108			
				81	162			

Notice in this scale,

a. The rows are in the ratio of 2:1.
b. A sequence with ratios of 3:1 runs down the lower left leaning diagonal of the Table.
c. The left leaning columns are in the ratio 3:2.
d. The right leaning columns are in the ratio 4:3.
e. Any number in this sequence is the arithmetic mean of the two numbers that brace it from above, e.g., 9 is the arithmetic mean of 6 and 12.
f. Any number of this sequence is the harmonic mean of the two numbers that brace it from below, e.g., 8 is the harmonic mean of 6 and 12 where harmonic mean is defined below.
g. Any integer from this sequence is the geometric mean of two numbers that frame it along any diagonal, e.g., 12 is the geometric mean of 6 and 24; 8 and 18; 9 and 16; etc.

Note: The arithmetic mean, c, of a and b satisfies the relation: $b - c = c - a$ or

$$c = \frac{a+b}{2}.$$ (21.1)

The harmonic mean, c, of a and b satisfies the relation: $\frac{b-c}{b} = \frac{c-a}{a}$, or

$$c = \frac{2ab}{a+b}. \tag{21.2}$$

The geometric mean c, of a and b satisfies: $a : c = c : b$ or $c^2 = ab$, or

$$c = \sqrt{ab}. \tag{21.3}$$

How does this table of integers relate to music and architecture? R. Witcover described in his book, Architectural Principles in the Age of Humanism, how Alberti used neo-Platonic ideas to create a system of architecture in which he considered a hexagon of integers surrounding an integer of the Nicomachus' Table 21.1. He then made adjacent integers the lengths, width, and heights of the rooms of his buildings or their facades [Wit].

Figure 21.2 shows the facades of two of Alberti's great buildings: Santa Maria Novella in Florence and San Sebastiani in Mantua, Italy. Although this system had harmony due to the many repeated ratios, it lacked additive properties. The Modulor of Le Corbusier, described in Fig. 18.4, and repeated in Table 21.2, and the Roman System of proportions described in Table 20.2a and repeated in Table 21.3, have all of the mathematical qualities of Alberti's system while also possessing additive properties.

(a)   (b)

Fig. 21.2.   (a) Santa Maria Novella, (b) San Sebastiani, Mantua

21.3. Modulor of Le Corbusier

We have seen in Chap. 18 on the Modulor of Le Corbusier that the Red and Blue sequences have many of the properties of Alberti's system. These scales are illustrated as follows (not drawn to scale):

Table 21.2. The Red and Blue Series of Le Corbusier

Blue		$\frac{2}{\phi}$	2	2ϕ	$2\phi^2$	$2\phi^3$
Red		1	ϕ	ϕ^2	ϕ^3	ϕ^4

where $\phi = \frac{1+\sqrt{5}}{2}$, the golden mean, and the Red and Blues sequence are both Fibonacci sequences in the sense that each term is the sum of the previous two terms, and where terms from the blue sequence are twice terms from the Red.

The Blue and Red sequences intersperse themselves so that a term from the Red divides the pair of Blue terms that brace it from above in the arithmetic mean while a Blue term divides the pair of Red terms from below in the harmonic mean, and each sequence is geometric echoing the properties of the Nicomachus Table 21.1. The sequence can also be expressed by integers where the Red sequence is a Fibonacci F-Sequence (see sequence (21.4)):

$$1 \quad 1 \quad 2 \quad 3 \quad 5 \quad 8 \quad 13 \quad 21 \quad 34\ldots \qquad (21.4)$$

and the Blue sequence is its double. The ratio of successive terms of the F-sequence approaches the *golden mean,* $\phi = \frac{1+\sqrt{5}}{2}$, in a limiting sense.

21.4. The Roman System of Proportions

The key to the Modulor system is the Fibonacci sequence which results in many additive properties. The engine behind the Roman system of proportions is the integer Pell sequence:

$$1 \quad 2 \quad 5 \quad 12 \quad 29 \quad 70 \quad 169\ldots \qquad (21.5)$$

Each term is twice the previous term added to the term before. Just as the ratio of terms of the Fibonacci sequence approaches the golden mean, the ratio of terms of the Pell sequence approaches nearer and nearer to the silver mean, $\theta = 1 + \sqrt{2}$.

The Roman System, described in Chap. 20, was based on an infinite number of Pell sequences in the sense that each term along a horizontal line is a Pell sequence shown in Table 21.3

Table 21.3.   The Roman System of proportions

...	$\dfrac{2}{\theta}$	2	2θ	$2\theta^2$	$2\theta^3$	...
...	$\dfrac{\sqrt{2}}{\theta}$	$\sqrt{2}$	$\theta\sqrt{2}$	$\theta^2\sqrt{2}$	$\theta^3\sqrt{2}$	...
...	$\dfrac{1}{\theta^2}$	$\dfrac{1}{\theta}$	1	$\theta \ \theta^2$	θ^3	...

Again each term intersperses the pair of terms that brace it from above, from below, and to the diagonals in the arithmetic, harmonic, and geometric means.

21.5.   Music and Proportions

Architecture articulates patterns in space while music articulates patterns over time. As a result, architecture has been likened to "frozen music." In architecture it is the ratio of its elements, or proportions, which is important rather than the absolute values of the lengths. The same is true in music. It is the relative frequencies, that matter, not the absolute pitch of the notes. Since music and architecture share common ideas, we will describe some of the rudiments of the musical scale. We have already seen that the Nicomachus Table 21.1 is constructed from the ratios: 2:1, 3:2, and 4:3.

How does the Nicomachus Table 21.1 relate to music? Notice that the boundary integers of the table are powers of 2 and 3, repeating the integers from Platos's "World Soul" (see Fig. 21.1). It is considered to be the miracle of music that when the relative frequency of a tone is multiplied by a power of 2, referred to as an *octave*, it sounds identical to the ear. We will show in Sec. 21.7 that the relative frequencies of all the tones of the musical scale, using a scale attributed to Pythagoras, can be generated by powers of the number 2 and 3 [Kap2].

Now consider a monochord, a string stretched between two bridges, where the pitch of the tone when sounded is related to the entire length of the string as shown in Fig. 21.3. I refer to this reference tone as the *fundamental*. Next place the bridge at the 1/2-half point of the string and sound the pitch. The result will be one octave higher than the fundamental. Now put your finger on the 2/3 point of the string and sound the tone. It will give rise to a pitch higher than the fundamental by a *perfect fifth*; i.e., if the fundamental tone is assigned the value D, the tone a fifth higher is A (DEFGA, five letters higher). Place the bridge at the 3/4 point and you will get the tone G, a *perfect fourth* (DEFG, four letters) higher. The inverse of the relative string length is the relative frequency. In what follows, we will express all tones in terms of their frequency. The Nicomachus Table 21.1 has been rewritten as Table 21.4. Just as with the integers in the Nicomachus Table 21.4 the musical scale is based on the proportions: 1:1, the unison; 2:1 , the octave; 3:2, the perfect fifth; and 4:3 the perfect fourth, the musical proportions recognized by Alberti. Also notice that column 5 of Nicomachus Table 21.4 contains five

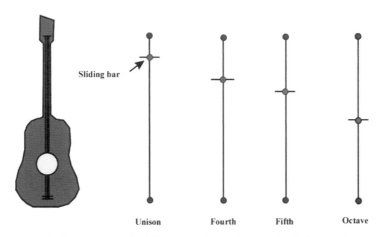

A sliding bridge on a monochord divides the string length representing fundamental tone into segments corresponding to musical fifth (2:3), fourth (3:4), and octave (1:2).

Fig. 21.3.   Segments corresponding to perfect fifths (2 : 3), fourths (3 : 4) and octaves (1 : 2)

successive perfect fifths which are the relative frequencies of the tones of the *pentatonic scale* while column 7, with seven successive musical fifths, are the tones of the *heptatonic scale*. These elements persist to this day and are related to what musicians refer to as tonic, subdominant and dominant chords which form the basis of Western music.

To create a musical scale within a single octave, a tone is chosen and referred to as the fundamental and assigned relative frequency 1 unit. The octave limit has relative frequency 2. Given any tone of the musical scale, we can always multiply its relative frequency by a power of 2 to bring that tone into the, [1, 2] octave interval. For example, given a relative frequency 3, the tone can be placed in the [1, 2] octave by dividing by 2 to get 3/2 which we have seen is identified with the perfect fifth. The tone with relative frequency 1/3 can be placed in the [1, 2] octave by multiplying by 4 to get 4/3 which we recognize as a perfect fourth.

In ancient times, 2 was taken to be the first female number while 3 was the first male number. In terms of music, 2 was thought of as "home", or the "great mother" yielding the same tone in different octaves when multiplied by the frequency of the tone. We have now seen that to get a new tone requires the frequency of the fundamental tone to be multiplied or divided by powers of the number 3, e.g., (3:2 or 4:3) and so male was associated with the creative element of music. Of course, these paternalistic symbols, although quaint, no longer hold in modern culture.

Table 21.4.   Nicomachus and the Musical Scale

1	2	4	G	8	16	C	32	64	F	...
	3	6	D	12	24	G	48	96	C	...
		9	A	18	36	D	72	144	G	...
			27	54	A	108	216	D		...
				81	E	162	324	A		...
					243	486	E			...
						729	B			...

In the next section we will describe the twelve tone *chromatic* scale of music and in Sec. 21.7 we will show how the entire musical scale can be created from the numbers 2 and 3.

21.6. The Musical Scale

Just as geometric proportions were at the basis of the three systems of architectural proportions described in this chapter, a geometric sequence spanning the octave between relative frequencies 1 and 2 also lies at the basis of the musical scale. The octave is divided into twelve equal proportional units corresponding to the twelve tones of the so called *chromatic scale*, as follows:

$$1, \ 2^{\frac{1}{12}}, \ 2^{\frac{2}{12}}, \ 2^{\frac{3}{12}}, \ 2^{\frac{4}{12}} = 3\sqrt{2}, \ 2^{\frac{5}{12}}, \ 2^{\frac{6}{12}} = \sqrt{2},$$
$$2^{\frac{7}{12}}, \ 2^{\frac{8}{12}}, \ 2^{\frac{9}{12}}, \ 2^{\frac{10}{12}}, \ 2^{\frac{11}{12}}, \ 2 \tag{21.6}$$

The numbers in this geometric sequence represent the relative frequencies of the 12 tones of the *chromatic* scale which increase in frequency like compound interest compounded at approximately 6% per year doubling after 12 years. Because the chromatic scale is expressed as a geometric sequence, following Chapter 15 it can be represented by a spiral. In Appendix 21.A, the twelve tones of the chromatic scale are shown to have a logarithmic nature (See Appendix 15.A).

These twelve tones can be represented as equally spaced points on a *tone circle*, as shown in Fig. 21.4, like a clock with the *fundamental tone* at 12 o'clock. Each interval is called a *semitone* while a pair of intervals is known as a *wholetone*. Note that the interval of the tritone from D to A flat has a relative frequency of $\sqrt{2}$, and it is often considered to be the most dissonant of intervals. However, each tone represents a pitch class of tones with the same tone in different octaves represented by the same point on the tone circle.

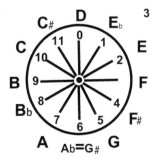

Fig. 21.4. The tone circle

If we go around the circle in a clockwise direction the tones increase in frequency. If we move in a counterclockwise direction the pitch of the tones decrease in frequency. As a result, movement around the circle should really be represented by a helix. In fact, it has been suggested by the ethnomusicologist, Ernest McClain, that all of the serpent myths of ancient times, where the serpent is either a spiral or helix, have their origins in the musical scale.

You will note that these tones are represented by irrational numbers rather than simple rational numbers such as 3/2 or 4/3. This is because the Chromatic Scale (21.6) is the *modern-equal-tempered* scale which was created to accommodate the invention of the piano which would require an infinite number of keys if the ancient system of rational numbers were used to express the tonal frequencies. However, the two scales are very close in their values. When D is the fundamental, the interval of the perfect fifth from D to A (DEFGA) is at 7 o'clock on the tone circle, and the perfect fourth from D to G (DEFG) is at 5 o'clock. Note that $\frac{3}{2} \approx 2^{\frac{7}{12}} = 1.498$ and $\frac{4}{3} \approx 2^{\frac{5}{12}} = 1.334$ so that the equal-tempered scale is a good approximation to the ancient scale of Pythagoras. You can also see that the interval from D to G can also be considered to be a falling fifth. Also A lies at a relative frequency of 3:2, the arithmetic mean of the octave [1, 2], while G lies at the harmonic mean 4:3 of the octave. The musical unit of cents is discussed in the Appendix 21.A to this chapter. In terms of cents, there are 1200 cents in an octave where each semitone amounts to 100 cents. Using cents, the rising fifth amounts to 700 cents on the piano while the falling fifth at 500 cents on the piano. Using the Pythagorean scale, the rising and falling fifths are 702 cents and 498 cents respectively.

In Fig. 21.5 the twelve tones of the chromatic scale are L spanned by the 88 keys of the piano. You will notice that the pattern of twelve tones repeats in different octaves. The white keys are represented by the letters: ABCDEFG while the black keys are augumented by sharps or flats. The scale of Western music is based on a sequence of seven tones. This heptatonic scale always begins on the fundamental which can be chosen to be any one of the twelve tones. Depending on the fundamental tone, the scale will be in a different *mode*. Successive white keys give rise to seven of the modes on the piano in Fig. 21.5. The most familiar mode to Western ears is the one beginning on C: CDEFGABC which is the famous: "do re mi ... scale". The mode beginning on D, or the ancient Phrygian mode, was the preferred mode of the ancient Greeks because the seven tones, DEFGABCD, are symmetrically spaced around the tone circle reflecting the symmetry of the pattern of keys on the piano as you can see by looking at Fig. 21.5, where the keys of the piano are symmetric about D. The three tones GDA are also shown in which the interval from D to A is a rising fifth while from D to G is a falling fifth.

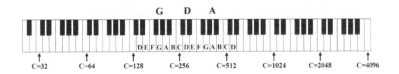

Fig. 21.5.   The piano keyboard

The three tones GDA basic to the musical scale are shown in Figs. 21.5 and 21.6a. Starting at the fundamental D, count 7 tones in a clockwise direction to the musical fifth at A and seven tones counterclockwise to the falling fifth at G.

The pentatonic scale is shown in Fig. 21.6b with the filled in circles representing the tones. The pentatonic scale with D as the fundamental is achieved by starting at D and counting off 7 tones in a clockwise direction on the tone circle to get the musical fifth at A, then counting 7 more clockwise tones to E. Now reverse field and count 7 tones in a counterclockwise direction, to G and seven more tones to C.

To create the heptatonic scale simply move 7 tones three times in a clockwise direction and three times in a counterclockwise direction as in Fig. 21.6c. Note that the open circles in Fig. 21.6b represent the black keys on the piano and so when Fig. 21.6c is turned upside down, you can see that the open pitches form the pattern of the pentatonic scale.

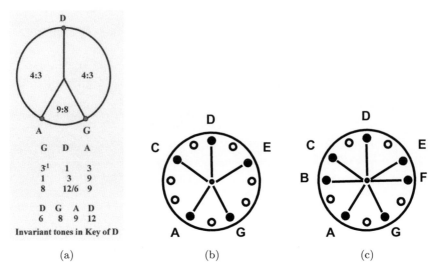

Fig. 21.6. (a) The three basic tones: GDA on the tone circle, (b) Tones of the pentatonic scale on the tone circle, (c) Tones of the heptatonic scale on the tone circle

Remark: The pentatonic and heptatonic scales have a striking difference. The pentatonic scale has no semitone intervals while the heptatonic scale has two such intervals. Semitone intervals are highly dissonant. As a result, a pentatonic scale sounds pleasant to the ear no matter how the tones are played. In other words, you will play nice sounding tunes by randomly sounding the black keys on the piano. This is why most of the folk music of the world is pentatonic. On the other hand, to play a piece of music in a heptatonic scale requires musical skill.

Exercise: If you have access to a piano, compose your own pleasant sounding piece of music by pressing on only the black keys of the piano in any order.

21.7. Creating the Chromatic Scale from the Numbers 2 and 3

The musical scale attributed to Pythagoras with integer values of the relative frequency based on the integer 3 is described in this section.

The ethnomusicalogist, Ernest McClain, provided the information for this approach [McC]. I use the following step-by-step approach:

a. The 12 tones of the equal-tempered scale are equally spaced around a tone circle like hours on a clock (see Fig. 21.4) with D atop the circle as the fundamental tone and the other tones in the octave between relative frequencies 1 and 2.

b. Each tone on the tone circle stands for a pitch class of tones differing by some number of powers of 2.

c. The relative pitch corresponding to integer 3 lies on the tone circle near 7 o'clock at A and can be obtained by counting 7 semitones clockwise around the tone circle, a perfect fifth. Although 3 does not fall in the [1, 2] octave, relative frequency 3/2 does. Using a system of cents as a measure of relative frequency, the fifth at 7 o'clock has 700 cents whereas relative frequency 3/2 has 702 cents, very close to the equal-tempered value. The measure of relative frequency in terms of cents is described in Appendix 21.A.

d. Two tones symmetrically placed on the tone circle have inverse relative frequencies. For example, the tone with relative frequency 1/3 is symmetrically placed with respect to 3 (also 3/2) at 5 o'clock or G. This tone can be thought to be obtained by moving 7 spaces in a counter-clockwise direction from D to G, a falling perfect fifth. It can be placed in the [1, 2] octave by its pitch class relative, 4/3. In terms of cents it has a relative frequency of 498 cents very close to the equal-tempered value at 500.

e. We have derived the rising and falling fifths and the fundamental: 1/3, 1, 3. We can relate this immediately to the 3rd column of the nicomachus Table 21.4 with explanation to follow,

Table 21.5.   The tetrachord

G	D	A
1/3	1	3
1	3	9
4	6	9
8	6/12	9

Row 1: The names of the tones of a rising and falling perfect fifth and the fundamental.

Row 2: The tones written as powers of 3.

Row 3: The tones are expressed as integers, done by multiplying by the common denominator, 3.

Row 4: Multiples of the integers in row 3 by powers of 2 yield the ascending frequencies of column 3 of Nicomachus Table 21.4 as a sequence of perfect fifths, i.e., 2^2, $2^1 \times 3$, $2^0 \times 3^2 = 4, 6, 9$. Note: multiplying by powers of 2 do not alter pitch class.

Row 5: Since the largest value of Row 4 is 9, and D is the fundamental, 9 is enclosed by the 6/12 octave while the other tone, 4, is multiplied by 2 to also place it in the 6/12 octave as 8.

Now that the relative frequencies, in a single octave (6/12), of the tetrachord G, D, A, and D' have been determined, they can be placed in scale order to get,

Table 21.6.   The tetrachord in scale order

D	G	A	D'
6	8	9	12

also visualized on the tone circle in Fig. 21.6a.

By using a similar technique, each column of the Nicomachus Table 21.4 can be placed in scale order.

f. Next, we apply this idea to creating a pentatonic scale summarized in Table 21.7:

Table 21.7.   The Pentatonic Scale

C	G	D	A	E
$\dfrac{1}{3^2}$	$\dfrac{1}{3}$	1	3	3^2
1	3	9	27	81
16	24	36	54	81
128	96	72/144	108	81

D	E	G	A	C	D'
72	81	96	108	128	144

We thus derive a pentatonic scale within a single octave. Again, notice that the sequence of integers in the 5th column of the Nicomachus table can be written as follows,

$$16 = 2^4, \ 24 = 3 \times 2^3, \ 36 = 3^2 \times 2^2, \ 54 = 3^3 \times 2, \ 81 = 3^4$$

showing how the terms depend on powers of 3.

g. Using the same reasoning results in one octave of the ancient Phrygian mode of the heptatonic scale,

Table 21.8. The heptatonic scale

D	E	F	G	A	B	C	D′
432	486	512	576	648	729	768	864

h. Twelve successive perfect fifths reach every tone on the tone circle as shown in Fig. 21.7 Another table illustrating these 12 fifths is given as follows:

Table 21.9. The 12 tone chromatic scale

A flat......	D......	G s
$\dfrac{1}{3^6}$	1......	3^6
$\dfrac{1}{2^{19}} = 524288$	$\dfrac{3^6}{3^{12}} = 531441$	3^{12}

On the piano 12 successive perfect fifths (12×7 tones) equals exactly seven octaves (7×12 tones). In other words, if you begin on a C and count off 7 keys (a fifth) 12 times you arrive at a C seven octaves above the starting C, i.e., For the Pythagorean scale this equality becomes an approximation, where $3/2$ is the relative frequency of a perfect fifth.

$$\left(\frac{3}{2}\right)^{12} \approx 2^7 \ \text{ or } \ \text{A flat} = 524288 = 2^{19} \approx 3^{12} = 531441 = \text{G sharp}.$$

$$(21.7)$$

We see that A flat and G sharp have pitches very close to each other. In fact, on the piano they would be the same pitch. The ratio of these tones is $r = \frac{531441}{524288} = 1.01364 = 23.45$ cents. (See Appendix (21.A) for a translation of relative frequency to cents.) Since each interval of a semitone on the tone circle amounts to 100 cents this ratio equals slightly less than a quarter of a semitone, referred to as the *Pythagorean comma*(see Sec. 21.6). Figure 21.8 shows that six rising fifths overshoots 6 o'clock on the tone circle while six falling fifths undershoot 6 o'clock on the tone circle.

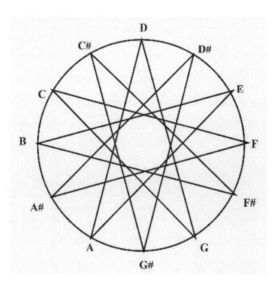

Fig. 21.7.   The cycle of perfect fifths

Remark: The "circle of fifths" is shown in Fig. 21.7 in which each tone of the chromatic scale results from a succession of 12 perfect fifths. This is sometimes referred to as the scale of spiral fifths. The double geometric sequence of 12 tones are therefore placed in pitch classes based on the number 3:

$$\frac{1}{3^6} \cdots \frac{1}{3^2} \frac{1}{3} \; 1 \; 3 \; 3^2 \cdots 3^6 \tag{21.8}$$

and according to Chap. 15, geometric sequences can be represented by a spiral (see Chap. 15).

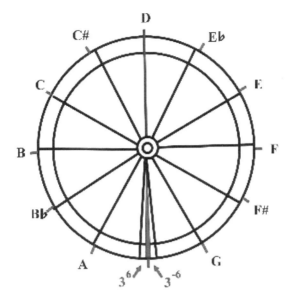

$G\# \neq A\flat$ **(Pythagorean = 729/512)**

or a $\flat \neq$ g# (Just = 45/32)

$G\# = A\flat$ **(Equal temperament =** $\sqrt{2}$ **)**

(Just = 45/32 = 760/512)

Fig. 21.8.   Illustration of the Pythagorean comma

21.8. The Relationship Between the Pythagorean Comma and the Solar and Lunar Years

In ancient cultures, 360 days was considered to be what was called the "canonical year," created for ceremonial purposes even though these societies were quite aware that 360 days was neither the length of the solar year of 364.25 days nor the lunar year of 354 days. However, the ratio of 360 days to the solar year and to the lunar year turns out to be remarkably close to the Pythagorean comma. For example,

$$\frac{364.25}{360} = 1.0118 = 20.309 \text{ cents.}$$

Is smaller than the Pythagorean comma by about 3 cents.

While, the ratio of lunar year to the canonical year is:

$$\frac{360}{354} = 1.0169 = 29.01 \text{ cents}$$

larger than the Pythagorean comma by about 5 cents.

The average of the solar and lunar years is then 24.66 cents, off from the Pythagorean comma by about 5%.

21.9. Additional Comments on the Musical Scale

1. It was a recent discovery that the various modes of the heptatonic scale, using the white keys of the piano, give expression to all of the patterns of African rhythms [Tou1,2]. If you clap loudly on the darkened points of the D or ancient Phrygian mode shown in Fig. 21.6b and clap softly on the empty points, you will get one of the clapping patterns. If you rotate the tone circle so that C is in the position of the fundamental at 12 o'clock, you will have the clapping pattern corresponding to the C mode. Through additional rotations you are able to derive all of the known African clapping patterns.

2. Another interesting finding was made by Anne Bulckens, who wrote her doctoral thesis on the proportions of the Parthenon, Ernest McClain and I joined Bulckens in a comprehensive study of the proportions of the Parthenon. The major result of our study is that the five most significant proportions of this great temple were the exact integers found in the fifth column of the Nicomachus Table 21.4. In other words, they were the same numbers that expressed the relative frequencies of the pentatonic scale when using a natural unit introduced by Bulckens that she called the dactyl. The dactyl is approximately the length of a joint of your finger. To be exact, 81 and 36 modules were the length and width of the outer temple where a module equaled 40 dactyls; 54 and 24 modules were the length and width of the inner temple or "cella"; and 16 modules was the height of the Parthenon to the entablature. Without using the unit of the dactyl, the relationship between the proportions of the Parthenon and the tones of the pentatonic scale would not have been detected. What is

interesting to note is that Bulckens had no knowledge of the musical scale when she did her work and so was not prejudiced in bringing forth this connection.

3. I also did some work with the Russian biophysicist, Sergey Petoukhov, who has discovered a strong mathematical system at the basis of RNA and DNA. The fourth column of the Nicomachus Table 21.4 contains the exact integers of the hydrogen bonds of the amino acids, the building blocks of RNA and DNA [Kap7]; I also recognized that the fourth column of the Nicomachus Table was a sequence of musical perfect fifths related to the hydrogen bonds of the amino acids.

4. The astronomer, Gerald Hawkins, related to me that the spacing of the moons of Jupiter and Saturn were also a sequence of musical fifths.

All of these connections point to the Greek understanding of human knowledge as expressed in the Quadrivium that every student should be advised to study number, geometry, astronomy and music and focus on unity as the basis of their education.

Appendix 21.A Logarithms and the musical scale

I suggest that you review the properties of logarithms in Appendix 15.A.

The relative frequency, r, of each of the twelve tones of the equal-tempered scale from the fundamental tone $r = 1$ to the octave value $r = 2$ is given by the geometric sequence,

$$1, 2^{\frac{1}{12}}, 2^{\frac{2}{12}}, 2^{\frac{3}{12}}, 2^{\frac{4}{12}}, 2^{\frac{5}{12}}, 2^{\frac{6}{12}}, 2^{\frac{7}{12}}, 2^{\frac{8}{12}}, 2^{\frac{9}{12}}, 2^{\frac{10}{12}}, 2^{\frac{11}{12}}, 2. \qquad (21.A.1)$$

Taking $\log_2$ of Sequence (21.A.1) results in the sequence of intervals,

$$0, \frac{1}{12}, \frac{2}{12}, \frac{3}{12}, \ldots, \frac{11}{12}, 1. \qquad (21.A.2)$$

Since the product of tonal ratios is equivalent to the sum of their intervals, the logarithm can be used to measure intervals since logarithms also have this property. For example, the relative frequency $r_1 = 2^{\frac{7}{12}}$ represents a perfect fifth as we saw in Sec. 21.7 while

$r_2 = 2^{\frac{5}{12}}$ represents a perfect fourth. The product of a fifth and a fourth is: $r_1 \times r_2 = 2^{\frac{7}{12}} \times 2^{\frac{5}{12}} = 2^{\frac{12}{12}} = 2$, the octave. Alternatively, using the logarithmic scale, $\log r_1 + \log r_2 = \frac{7}{12} + \frac{5}{12} = \frac{12}{12} = 1$ which states that the number of intervals of a fifth and a fourth result in the octave. We can now state the general principle of the musical scale.

General Principal of Music: As the relative frequencies of a pair of tones multiply their intervals add

In order to give each interval of the equal-tempered scale a value of 100 cents, we multiply the logarithm of the relative frequency of each tone in Sequence (21.A.2) by 1200 so that each of the 12 semi-tones within an octave is assigned a multiple of 100 *cents*. Since the sequence of intervals is logarithmic, to convert a relative frequency r to cents, the following formula can be used,

$$r_{cents} = 1200 \log_2 r = 1200 \times 3.322 \log_{10} r \qquad (21.A.3)$$

where 3.322 is the conversion factor between $\log_2$ and $\log_{10}$ (see Appendix 15.A.).

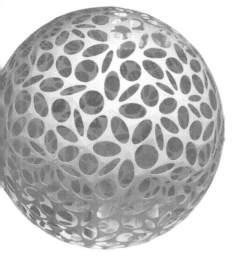

Chapter 22

Zonogons

22.1. Transformation of Regular Tilings

Starting with a polygon or aggregation of polygons in the form of a tiling of the plane, and applying a set of rules of transformation, there are several ways in which new polygons or tilings of the plane can be generated. Robert Williams [Wil] has considered a number of different classes of transformations. In this chapter we will consider three of these classes, namely, distortions and augmentation-deletion.

22.2. Distortion

The distortion operation consists of expanding, contracting, twisting, flattening, and stretching polygons either in isolation or in aggregation. One special type of distortion operation involves n-zonogons. An n-zonogon is a $2n$-sided polygon with opposite sides equal and parallel. For example, the parallelograms and hexagons that combine to tile the plane regularly are 2-zonogons and 3-zonogons respectively. An n-zonogon can be constructed by specifying what is referred to as an n-vector star of n-directed line segments (vectors) representing the orientation and length of its sides. For example, a star of 3 vectors and the resulting 3-zonogon is shown in Fig. 22.1a and 22.1b.

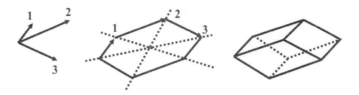

Fig. 22.1.　A 3-vector star results in a hexagon and a parallelepiped

Also, zonogons have the property of being centrally symmetric as shown in Fig.22.1b. In addition, an n-zonogon can always be decomposed into $n(n-1)/2$ parallelograms, the total number of ways in which two vectors can be chosen from a set of n vectors when order is not important. This is demonstrated in Fig. 22.1c for a 3-zonagons in two different ways.

22.3.　Augmentation-Deletion

What is important about zonogons is that they can be contracted or expanded in a direction parallel to a pair of opposite sides as shown in Fig. 22.2, without altering the angles between adjacent sides. Thus, any space filling aggregate of zonogons will remain space filling after distorting an individual zonogon in this way, and then adjusting adjacent zonogons of the tiling accordingly, as shown in Fig. 22.3.

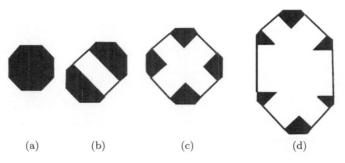

(a)　　　　　(b)　　　　　(c)　　　　　(d)

Fig. 22.2.　Distortion of an octagon in three directions

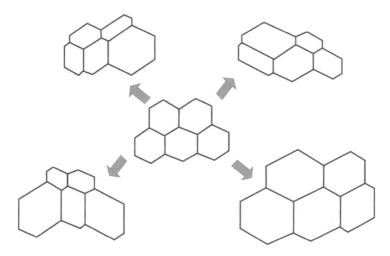

Fig. 22.3.   Distortion of a tiling by hexagons

Construction: Construct an interesting aggregate of 3-zonogons starting with a vector star of 3 vectors of your choosing. Your aggregate should illustrate the ability to fill space.

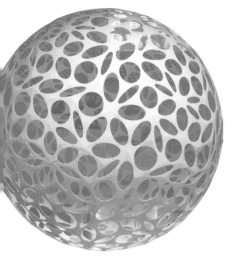

Chapter 23

Tangrams and Amish Quilts

The artist is like Sunday's child; only he sees spirits. But after he
has told of their appearing to him everybody sees them.

— Goethe

23.1. Introduction

While driving with my family on a vacation in Lancaster County, the
home of the Pennsylvania Dutch, I began to make plans for my course
on the Mathematics of Design. I wanted to find a way of linking ideas
from the history of design to the world around me.

I had just been reading *Secrets of Ancient Geometry* by Tons
Brunes [Bru] in which he analyzes an enigmatic eight-pointed star
that was the subject of Chap. 6. He describes his theory that this star,
along with the subdivision of a square by a geometrical construction
that he calls the "sacred cut", formed the basis of temple construction
in ancient times (see Chap. 20). That led to the Roman system of
proportions and to the construction of the octagon as described in
Chap. 20.

Kim Williams, an architect living near Florence, also described
to me how she had found the system related to Brunes' sacred-
cut geometry embedded in the proportions of the pavements of the
Baptistry of the church of San Giovanni which itself is shaped like
a regular octagon [Wil]. The pavements themselves had many star

octagonal designs engraved in them. The star octagon, an ecclesiastical emblem, signifies resurrection. In medieval number symbolism, eight signified cosmic equilibrium and immortality.

23.2. Tangrams

Recently, I had been showing my son the fascinating *tangram puzzle,* shown in Fig. 23.1a in which thousands of pictograms, such as the one shown in Fig. 23.1b, are created from the dissection of a square into the seven pieces shown in Fig. 23.1a. A tangram set can be created from a single square piece of paper by simply folding and cutting. The pieces consist of a 45 deg. right triangle at three different scales along with the square and diamond formed by juxtaposing two 45 deg. right triangles as shown in Fig. 23.2. The side of the larger triangle is equal in length to the hypotenuse of the next smaller triangle. Each pictogram must be formed from all the seven pieces with no repeats and no overlaps. Enlarge the pieces, cut them out, and try your hand at constructing the pictogram shown in Fig. 23.1a. Exactly 13 convex polygons (polygons with no indentations) can be constructed from the tangram set including one rectangle (other than a square) and one triangle (other than an isosceles right triangle). However, it is enough of a challenge to reconstruct the square.

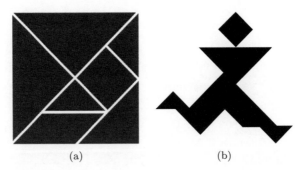

(a)   (b)

Fig. 23.1.   (a) The tangram set; (b) a pictogram constructed with the tangram set

Construction: Try to construct as many of the 13 convex polygons that you can from the Tangram set.

23.3. Amish Quilts

On our vacation to Pennsylvania Dutch country we were able to explore the country-side, visit working farms, and delve briefly into the rich history of the people. The Amish and Mennonites settled in Pennsylvania during the 18th and 19th centuries as refugees from religious persecution in Germany and found a haven of freedom and rich farm lands in Lancaster county. While the Mennonites are devoutly religious and live simple lives devoid of materialistic pursuits, they do enjoy a few of the comforts of modern society. The Amish, however, attempt to insulate themselves as much as possible from outside influences and live a plain existence in which they farm without electricity, drive horse drawn carriages, and wear unostentatious clothing. Amish women live extremely proscribed lives caring for the house and children. One of the few outlets for their creativity is the practice of quilt making [Ben].

Fig. 23.2.   The 45 deg. right triangle is the geometric basis of the tangram set

The oldest known quilts date to about 1850. However, quilting designs have changed only slightly through the years. Geometric patterns consisting of squares, star octagons, diamonds and 45 deg. right triangles are used in simple designs. While the geometric patterns are the manifest content of the quilts, fabric is stitched with a variety of subtle patterns such as tulips, feathers, wreaths, pineapples, and stars. I purchased a quilt with the design shown in Fig. 23.3a. I was amazed to see that it consisted almost entirely of pieces from

a tangram set. You can see that it has 45 deg. right triangles at
three different scales, as shown in Fig. 23.2, squares, and diamonds
that have the same internal angles as the tangram diamond, namely
45 deg. and 135 deg. However, the Amish quilt diamonds differ from
the Tangram diamonds by having all equal edge lengths. The ratio
of the diagonals of the Amish diamond is $(1 + \sqrt{2}) : 1$ which is an
important number referred to as the silver mean that was considered
in greater depth in Chap. 20. This is identical to the ratio of line
segments into which the sacred cut divides the edge of a square. I
shall refer to these as Amish diamonds.

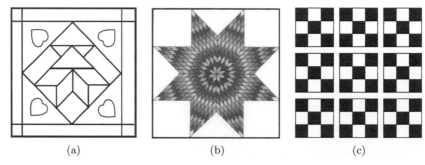

Fig. 23.3.   (a) A traditional Amish quilt illustrating the tangram pieces; (b) an
Amish quilt made up of "Amish diamonds"; (c) the Amish "nine-square" pattern

I also purchased a larger quilt which utilizes the pattern of the
star octagon shown in Fig. 23.3b. It is made of the Amish diamonds
in rhythmically alternating colors so that it appears to be pulsating
energy into a room. Finally, I purchased two potholders in the basic
Amish nine-square pattern (Fig. 23.3c). I was soon to discover that
the nine-square was intimately related to Brunes' star figure (see
Chap. 6). At last I had the connections that would give unity and
substance to my course.

Construction: Create your own Amish quilt design consisting of
polygons from the tangram puzzle and the Amish quilt patterns presented in this chapter. You can see several examples on the website.

I shall now summarize some of the geometric connections related
to these personal discoveries as I reported them to my class.

23.4. Zonogons

Zonogons are polygons with opposite edges parallel and equal [Cox], [Lal], [Kap2]. They were described in Chap. 22. The edges are oriented according to the directions given by a vector-star. An example of a 4-zonogon is the regular octagon which can be tiled with two squares and four Amish diamonds in two different ways, is shown in Fig. 23.4 (if the tangram diamonds are used, the octagon will not be regular). This is an example of a more general result that says that an n-zonogon can be tiled by $N = \frac{n(n-1)}{2}$ parallelograms in two distinct ways, e.g., for $n = 4$, by $N = 4 \times 3/2 = 6$ parallelograms. An n-zonogon is a polygon with $2n$ sides, i.e., a $2n$-gon with n pairs of parallel and congruent edges. The edges of its parallelogram tiling are oriented in n vector directions. This is illustrated in Fig. 23.4 for the 4-zonogon with its 4-vector star in 4 different vector directions. The central angle of the regular octagon is represented by $\theta = \frac{360}{8} = 45$ deg. in Fig. 23.5a, while the two different types of parallelogram tilings derived from the 4-zonogon are shown in Fig. 23.5b to have angles $(1\theta, 3\theta)$ and $(2\theta, 2\theta)$. We use the shorthand form $(1, 3)$ and $(2, 2)$ of this notation to represent the two distinct angles of a parallelogram (the other two angles are repeated). Notice that the angles add up to $4\theta = 180$ deg., (see Fig. 23.5) whereas the angles surrounding each vertex in Fig. 23.4 sum to 8θ, or 360 deg. This can easily be generalized to n-zonogons and their derived parallelograms

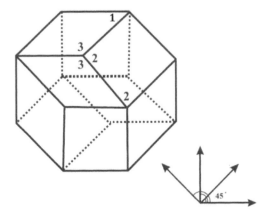

Fig. 23.4. A regular octagon tiled with two squares and four Amish diamonds in two ways

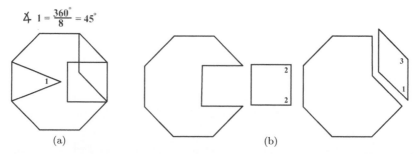

Fig. 23.5. (a) The parallelograms defined by a 4-zonogon; (b) the two angles of the parallelograms add to $4\theta = 180$ deg.

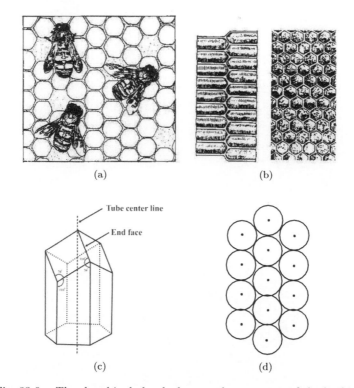

Fig. 23.6. The rhombic dodecahedron as the structure of the beehive

[Cox], [Lal2], [Lal3]. The tiling of the octagon by six parallelograms in two ways can be seen to be the projection of a polyhedron called the *rhombic dodecahedron* which is also related to the structure of a beehive shown in Fig. 23.6. This was a fitting end to my search for geometric forms related to nature to relate to my students.

A key property of n-zonogons is that the edges of the parallelograms that tile it line up in a series of n sets of parallel edges or zones. The edges of each zone are oriented in the direction of one of the n vectors that define the n-star of the zonogon. You can observe this in Fig. 23.4 for the 4-zonogon. If the length of one of the vectors shrinks to zero, then one of the zones is eliminated and the n-zonogon collapses to a $(n-1)$-zonogon. Alternatively, each of the n vectors can be expanded or contracted, with the effect that the shape of the zonogon is distorted without altering the internal angles of its parallelograms as in Chap. 22.

23.5. Zonohedra

The two sets of parallelograms that tile the 4-zonogon in Fig. 23.4 can be seen to be a projection of a twelve-faced, space filling polyhedron known as a rhombic dodecahedron (RD) [Kap 3].

The rhombic dodecahedron is representative of a class of polyhedra with opposite faces parallel and congruent known as zonohedra. Each three connected edge forms a parallelopiped where

$$C(n,3) = \frac{(n-2)(n-1)(n)}{6} \tag{23.1}$$

is the number of ways one can choose three objects from a group of n, where order is not important. If this is done, then n edges are incident at each vertex giving a projection of an n-dimensional cube in 2 or 3 dimensions. But what do we mean by an n-dimensional cube?

Let's consider a 4-dimensional cube, or tesseract as it is called, the boundary of which, in one of its two-dimensional projections, is a 4-zonogon. We see it pictured in Fig. 23.7 as the fifth in a series of 0, 1, 2, 3, and 4-dimensional cubes. The 0-dimensional cube (see Fig. 23.7a) is a point with no degrees of freedom. The surface of a 1-dimensional cube (line segment) is gotten by translating the 0-dimensional cube (point) parallel, to itself (see Fig. 23.7b). One has freedom to move left or right along the line. The surface of a 2-dimensional cube (see Fig. 23.7c) is gotten by translating the line segment parallel to itself to obtain a square. Movement is possible on the surface of the square: left-right or up-down. A 3-dimensional cube (see Fig. 23.7d) is obtained by translating a square parallel to itself,

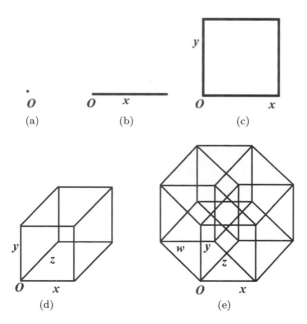

Fig. 23.7.   Diagrams of 0, 1, 2, 3, and 4-dimensional cubes

resulting in a surface with freedom of movement: left-right, up-down, in-out. Finally, the 4-dimensional cube (see Fig. 23.7e) is obtained by translating the 3-dimensional cube parallel to itself. You can see that now 4 degrees of freedom are possible: left-right (x), up-down (y), in-out (z), and movement in the elusive fourth direction (w). Of course, Fig. 23.7e is only the projective image of a 4-dimensional cube the same way that Fig. 23.7d is only a projection of a 3-dimensional cube. In an actual 4-dimensional cube there would be no intersecting lines, planes, or volumes just as a 3-dimensional cube has no crossing edges despite the crossing edges that appear in its 2-dimensional projection. Also the 4 directions are mutually perpendicular in 4-dimensions.

Notice the star octagon in Fig. 23.7e, reminiscent of my Amish quilt.

23.6.   Triangular Grids in Design: An Islamic Quilt Pattern

A 3-zonogon is shown in Fig. 23.8. The two sets of 3 parallelograms that tile the hexagon can be seen to be an ordinary cube

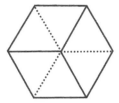

Fig. 23.8.   A 3-zonogon viewed as either a 3-dimensional cube or as a triangular grid

in perspective. The hexagon is also subdivided into a triangular grid. This triangular grid is useful as a design tool and was studied in depth in Chap. 1.

In Fig. 23.9b we see a triangular grid developed from a triangle-circle grid in Fig. 23.9a (see Chap. 1). Repeating patterns can be created by deleting edges from Fig. 23.9a. Two examples are shown in Fig. 23.10.

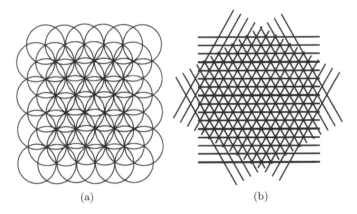

(a)   (b)

Fig. 23.9.   (a) A triangular grid of closely-packed circles; (b) a triangular grid

Margit Echols [Ech4] has developed geometrical principles suited to the particular requirements of the art of quilting. One of her quilts is based on an Islamic pattern generated from the triangular grids of Figs. 23.9a and 23.9b. Her quilt pattern, illustrated in Fig. 23.11, contains twelve pointed stars. Three pairs of bounding edges of the star, when extended, traverse the entire pattern and form a triangular grid. Notice that pentagonal star-like figures make a surprise appearance in the final design.

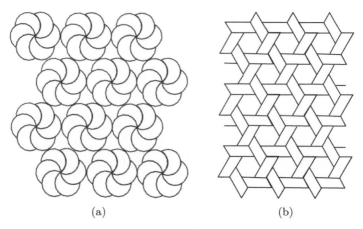

(a)   (b)

Fig. 23.10.   Two patterns conforming to (a) the grid of close packed circles; (b) the triangular grid

Fig. 23.11.   "Cairo Quilt" by Margit Echols 1994, Cotton, $90'' \times 110''$, machine pieced, hand quilted

Echols has the following to say about the art of quilt making:

"The quiltmaker is faced with tremendous restrictions inherent in both the laws of geometry and the technology of patchwork. How is it we can bear the time it takes to make a quilt? — Besides the obvious rewards of accomplishing technical challenges, of making colors sing, the tactile sensuality of textiles, and the meditative quality of repetitive handwork — there is the pleasure of problem solving of putting the puzzle together, of playing the game, a serious game of a battle against chaos which has deep intellectual appeal."

23.7. Other Zonogons

For a 5-zonogon, the central angle is $\theta = \frac{360}{10} = 36$ deg. and the two species of parallelograms are $(1, 4)$ and $(2, 3)$ in the shorthand notation, adding up to 5θ. These parallelograms have interesting properties since the ratio of their edge length to one of their diagonals is the golden mean, a number whose value is $\phi = \frac{1+\sqrt{5}}{2} = 1.618\ldots$ (See Chap. 16). These parallelograms were discussed in Sec. 19.3. Designs with these parallelograms, such as the one in Fig. 23.12, have approximate five-fold symmetry.

Construction: Construct 4, 5, and 6-zonogons, and tile them with parallelograms in two ways. You may use color to make them into designs.

The design possibilities are all the richer for tiling a 6-zonogon. Tiling the 6-zonogon by its parallelograms: $(1, 5)$, $(2, 4)$ and $(3, 3)$ where $\theta = \frac{360}{12} = 30$ deg., results in perspective diagrams of the rhombic triacontahedron (30 parallelogram faces) and the truncated octahedron (with 6 square and 8 hexagon faces) shown in Fig. 23.13. Focusing on the rhombic triacontahedron, it is a zonohedron with all of its faces congruent $(3, 2)$ parallelograms. It is also the surface of the 2-dimensional projection of a 6-dimensional cube. Using the zometool kit, you can build a 3-dimensional model of the rhombic triacontahedron. Then by successively removing one set of zones, the 6-zonohedron (rhombic triacontahedron) collapses to a 5-zonohedron (rhombic icosahedron), then to 4-zonohedron (rhombic dodecahedron) and finally to a 3-zonohedron (parallelepiped). In the last step

Fig. 23.12.   A pattern with approximate five-fold symmetry made up of the two parallelograms of the 5-zonogon from the Mathematics of Design by Jay Kappraff

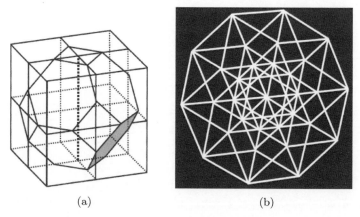

Fig. 23.13.   (a) The truncated octahedron; (b) the rhombic triacontahedron (Due to Lalvani)

we are left with two parallelopipeds with $(3, 2)$ faces depending on how the zones are removed. I refer to these parallelepipeds as golden parallelopipeds of Type 1 and Type 2 that are the building blocks of the zonohedra related to the 6-dimensional cube (see Fig. 23.14). All faces of this family of zonohedra are congruent $(3, 2)$ rhombuses that have diagonals in the ratio, $\phi : 1$, and for this reason they are referred to as *golden iso-zonohedra* [Miyazaki, 1980]. Each zonohedron can be tiled by the number of parallelopipeds given by Eq. (23.1). For example, the rhombic dodecahedron with $n = 4$ is tiled by 4 parallelopipeds, 2 of type 1 and 2 of type 2 as shown in Fig. 23.15. The rhombic triacontahedron, shown in Fig. 23.16, with $n = 6$, is tiled by 20 paralellopipeds, 10 of type 1 and 10 of type 2. It should be noted that these 20 paralellopipeds self-intersect in this 2-dimensional projection of a 6-dimensional cube, but are non-intersecting in 6-dimensions.

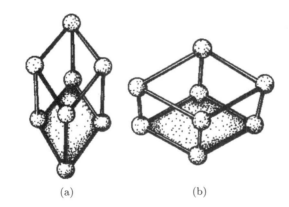

(a)   (b)

Fig. 23.14.   A golden parallelepipeds of type 1 and type 2

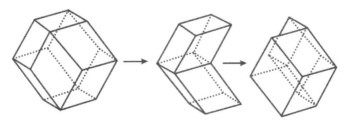

Fig. 23.15.   A rhombic dodecahedron tiled by two golden parallelopipeds of type 1 and two parallelopipeds of type 2

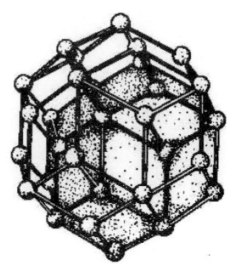

Fig. 23.16.   A rhombic triacontahedron tiled by golden parallelopipeds (by H. Lalvani)

It should be noted that in 1981 H. Lalvani generalized the Penrose tiling as the $n = 5$ case of n-dimensional cubes when he discovered the angle-sum rule: only permissible pairs of numbers that add up to n determine the shapes and number of possible rhombii from n-dimensions [Lal4].

The 6-zonogon can also be viewed as a distorted 2-dimensional projection of a 6-dimensional cube, and as for the 6-zonogon, it too has a star dodecagon (12 pointed star) at its center (see Fig. 23.17). We also encountered this star in Fig. 21.8 in connection with tone cycles of musical fifths. The cover of my book, Connections [Kap3], shows the extraordinary result of truncating a 6-dimensional cube at one of its vertices.

Alan Shoen [schoe] has created a puzzle called Rhombix in which multicolored tiles which are composites of 8-zonogons are used to create a prescribed set of designs. The artist, Tony Robbins [Rob] has built a 60-foot sculpture, shown in Fig. 23.18, based on quasicrystal geometry for the three story atrium at Denmark's Technical University of Copenhagen. The dome is a rhombic tricontahedron with a quasicrystal interior.

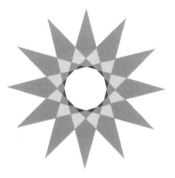

Fig. 23.17.   A star dodecagon

Fig. 23.18.   A 60-foot long rhombic triacontahedron sculpture with a quasicrystal interior at Denmark's Technical University in Copenhagen by Tony Robbin

In my book Connections, I described some morphological discoveries of Haresh Lalvani [Lal5] including a new class of hyper-geodesic surfaces. These are portions of tilings composed of rhombii, similar to the Penrose tiling, and are 2-dimensional projections from five and six dimensions. These tilings are mapped on various curved surfaces like spheres, cylinders, toruses, minimal surfaces and faces of polytopes in [Lal6]. A few years after that publication [Lal7] describes how Lalvani built the *HyperSurface* series of sculptures from flat metal sheets. Three examples from this series are shown in Figs. 23.19a, 23.19b, and 23.19c, a sphere, an elongated ellipsoid and an oblate ellipsoid. Part of the entire collection is shown in Fig. 23.19d from the Moss Gallery display at Design Miami, 2001. One of Lalvani's

Fig. 23.19. (a) HOLEYSPHERE (2007), painted steel, laser-cut, 24″ dia.;
(b) ALIEN (2008, '10), painted stainless steel, laser-cut, 60″ high, 36″ dia.;
(c) BLATE (2008), stainless steel painted, laser-cut, 17″ high, 36″ long;
(d) HYPERSURFACES display, Moss Gallery, Design Miami, 2011 by H. Lalvani

sculptures from this series can be seen on a street plaza in midtown Manhattan on 54th Street and Avenue of the Americas.

23.8. Conclusion

The concept of the zonogon is the key to understanding the system of design used by the Amish and illustrated by the tangram set. It leads to a means of visualizing the 2-dimensional projection of an important class of three-dimensional polyhedra known as zonohedra, and it places an understanding of the nature of 3-dimensional geometry firmly in the context of higher-dimensional geometry. The 4- and 5-zonogons define systems with a repertoire of two parallelograms, the first related to $\sqrt{2}$ and the sacred cut, the second related to the golden mean. A system of architectural proportions developed by the Le Corbusier, known as the Modulor (see Chap. 18), is based on the golden mean [Kap3].

In Chap. 20 we explored the $\sqrt{2}$ system of proportions in greater depth. We also saw that these two systems share a unifying structure with its roots in the musical scale. The number of parallelograms proliferate for zonogons of a higher order which inhibits their usefulness to serve as systems of proportion.

My visit to the Amish country, examination of the quiltwork of Margit Echols, the structures of Tony Robbin and the sculptures of Haresh lalvani have reinforced my feeling that artists, and practitioners of the folk arts have infused their work with patterns that share themes of common interest to mathematicians and scientists.

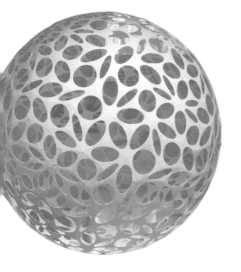

Chapter 24

Mirror Curves by Slavik Jablan and Ljiljana Radovic

The Tchokwe people of Angola play a game in which they drop a set of stones onto a bed of sand and try to encircle them by a curve drawn with a stick in the sand with a single stroke creating what are called Sona drawings. Some men from the village have reputations of being masters at playing this game. The ethnomathematician, Paulus Gerdes [Ger], studied this game and discovered that it was based on what are called mirror curves. This chapter will describe these mirror curves and their mathematics [Jab3, Jab4, Rad4, Jab7]. In the next chapter they will be used to create designs referred to as Lunda patterns.

We begin with a single square with a black dot in it (Fig. 24.1a). Additional squares form an $a \times b$ rectangular grid, RG$[a, b]$, where a and b are integers and 'a' is the number of squares from left to right while 'b' are the number of squares from bottom to top as shown in Fig. 24.1b. The square in Fig. 24.1a can be broken down to four sub-squares, called *fields* in which four sub-squares share a black dot and this spreads to the other squares in Fig. 24.1b, a 4×3 grid, RG$[4, 3]$, with $a = 4$ and $b = 3$. In the last step, the grid can be removed as in Fig. 24.1c. The entire grid is imagined to be surrounded by one-way mirrors.

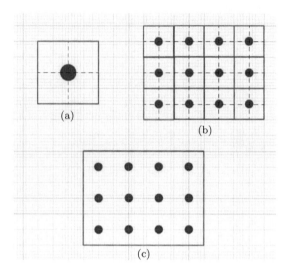

Fig. 24.1. (a) A single square with a black dot and four fields, (b) 4 × 3 grid of black dots, (c) 4 × 3 grid with the grid lines removed

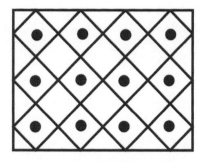

Fig. 24.2. A beam of light encircling the grid and surrounding the dots

We imagine that a beam of light is emitted from the midpoint of one of the squares on the edge of the grid reflecting from the mirror on another edge and moving along the diagonals of the sub-squares that make up the field in an oriented curve. This reflected beam of light encircles the grid surrounding the dots and returns to the starting point as shown in Fig. 24.2. Sometimes the light curve encircles all of the dots in a single stroke as in Fig. 24.2, but other times there are dots not all surrounded. In this case we start with a new ray of light emitted from another starting point and continue in this way until the whole of RG[a, b] is uniformly covered by mirror curves. You may need several curves to encircle all of the dots. Two-sided mirrors placed between cells, coinciding with internal edges (cell borders) or perpendicular to them at their mid-points, can help to facilitate the creation of a single or monolinear Sona drawing if placed properly in the grid. The grid may contain several mirror curves and your job is to insert the two-sided mirrors into the grid in such a way that a monolinear curve is obtained.

The next example shows different situations that can occur depending on the value of the greatest common divisor c of a and b for RG[a, b], i.e., $c = \mathrm{GCD}(a, b)$. You can show that the number of components (it is useful to represent them with different colors, so they can be easily distinguished) is equal to $c = \mathrm{GCD}(a, b)$. For example, in Figs. 24.3a and 24.3b, for $a = 2$ and $b = 2$ the number of components is $c = \mathrm{GCD}(2, 2) = 2$, for $a = 4$ and $b = 2$ the number of components is $c = \mathrm{GCD}(4, 2) = 2$, for $a = 5$ and $b = 2$ the number of components is $c = \mathrm{GCD}(5, 2) = 1$, for $a = 6$ and $b = 2$ the number of components is $c = \mathrm{GCD}(6, 2) = 2$, for $a = 6$ and $b = 3$ it is $c = \mathrm{GCD}(6, 3) = 3$, *etc.*

Notice in Figs. 24.3a and 24.3b the curves have rounded corners. In order to draw curves with smoothed, rounded corners, for reflecting a ray of light in the external edges and in the corners of the grid, you can use the following rules in Figs. 24.4a and 24.4b (this can be done in *Adobe Illustrator* by the command Filter > RoundCorners, or by command Effect > Stylize > RoundCorners).

Let's look in depth at an example of what will happen if the numbers a and b are not relatively prime. As before, we draw a rectangular grid, but now of dimensions 4×4, and we place the rays of light (components) in it. In order to surround all dots and uniformly cover RG[4, 4] by the mirror curve, we need four components since $\mathrm{GCD}(4, 4) = 4$ (Fig. 24.5).

The next question is how to place internal two-sided mirrors within the existing scheme, in order to join our components into a

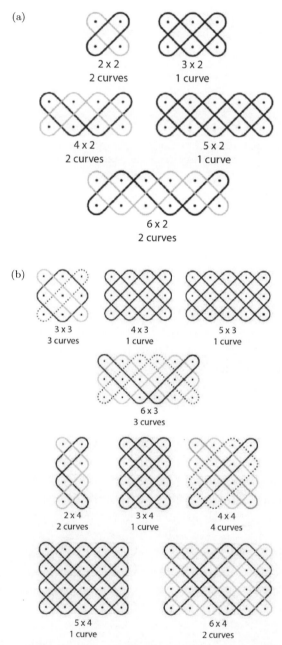

Fig. 24.3.   The number of curves encircling an $a \times b$ grid equals $\text{GCD}(a, b)$. (a) Examples: 2×2, 3×2, 4×2, 5×2 and 6×2 grids (b) Examples: 3×3, 4×3, 5×3, 6×3, 2×4, 3×4, 4×4, 5×4, 6×4 grids

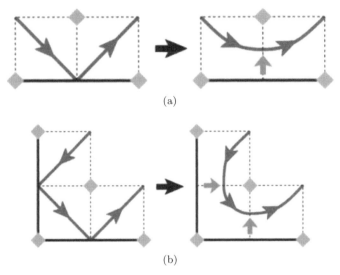

Fig. 24.4. (a, b) Creating smooth mirror curves with smooth and rounded corners

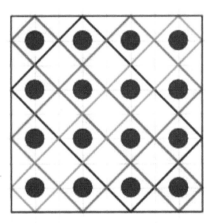

Fig. 24.5. A light ray through a 4 × 4 grid breaks into 4 mirror-curves

single mirror curve, i.e. to get a single ray of light that traces it. When inserting an internal mirror, we have two possible choices (Fig. 24.6):

(a) place it between square cells, incident to the internal edge;
(b) place it perpendicular to the edge at its mid-point.

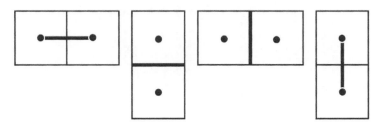

Fig. 24.6.   Internal mirrors placed midway between dots or joining adjacent dots

A mirror placed at the crossing point of two components (denoted by different colors) will join them into one. Placed at a self-crossing of an oriented component, depending on the position, a mirror either does not change the number of curves, or breaks the curve into two closed curves.

In the case of a rectangular square grid $RG[a, b]$ with sides a and b, the initial number of curves, obtained without using internal mirrors is $k = \text{GCD}(a, b)$ [Jab4]. If we want to transform it to a monolinear design, first introduce $k - 1$ internal mirrors at crossing points belonging to different curves. After that, when the curves are connected and transformed into a single curve, we may introduce other mirrors (for example, if we want a more symmetrical or specific design), taking care about the number of curves, according to the rules mentioned above. For example, see Fig. 24.7 and Fig 24.8. Notice that symmetrical placements of mirrors result in symmetrical mirror curves.

Fig. 24.7.   A 2×2 curve with internal mirrors

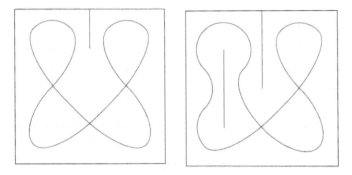

Fig. 24.8.   A smooth curve through a 2 × 2 grid with 1 and 2 internal mirrors

After inserting internal mirrors, we again make a smoothing of the resulting mirror curve. The next image shows the smoothing of zig-zag lines (Fig. 24.9).

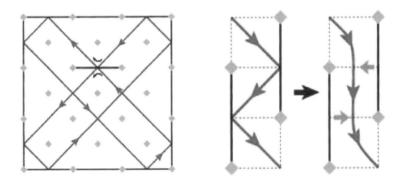

Fig. 24.9.   Smoothing a mirror-curve through a 3 × 3 grid

Let's return to the example of the 4 × 4 grid covered by four components. After inserting an internal mirror at the crossing point of two different curves (represented by different colors), the number of components reduces to three, since the two curves are joined into one (after this, you need to make a new coloring of the components, representing the joined curves by a single color). Continuing in this way by inserting new internal mirrors at the crossing point of two components, at each step the number of curves decreases by one, so that after three steps we obtain a single component — monolinear mirror curve uniformly covering RG[4, 4]. Figure 24.10 shows one possible way to introduce three internal mirrors in order to get a single curve. Certainly, the solution of this problem is not unique; you can choose different arrangements of three internal mirrors each resulting in a monolinear mirror curve.

In Fig. 24.10 the mirrors are not symmetric, so the resulting mirror curve is asymmetric as well. Figure 24.11 shows a symmetric arrangement of four internal mirrors according to a rotation of order

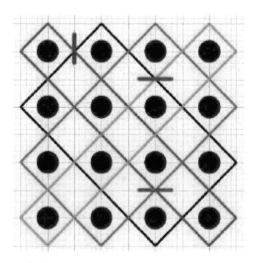

Fig. 24.10.   A curve through a 4 × 4 grid with 3 internal mirrors results in a mono-linear mirror-curve

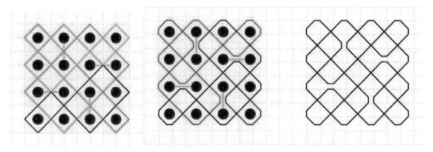

Fig. 24.11.   A symmetric arrangement of internal mirrors in a 4 × 4 grid results in a symmetric but not connected mirror-curve

4, so the resulting drawing has the cyclic rotational symmetry of the group C_4. However, we have not completely followed the rule about pairing components of different colors, and so we obtained a 4-component (imperfect-non-monolinear) mirror curve. Try to make another symmetrical arrangement of internal mirrors in the same — 4 × 4 grid resulting in a perfect (monolinear) mirror curve. Notice that symmetrically placed mirrors result in symmetric mirror curves.

Figure 24.12 illustrates the same exercise within a 3 × 3 grid. At the beginning, we have 3 components, so it is necessary to insert at least 2 internal mirrors in order to obtain a single curve. By using two internal mirrors, placed asymmetrically, we obtain the drawing shown on the left (Fig. 24.12a). The right portion of Fig. 24.12b shows the same curve turned into the diagram (projection) of an alternating knot, by introducing at every crossing the relation "over-under" in an alternating manner. You will find many similar examples in books describing Celtic knotworks or in computer programs for drawing Celtic knots.

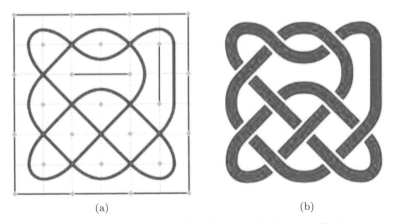

(a)   (b)

Fig. 24.12. (a) A curve through a 3 × 3 grid with 2 internal mirrors results in a single, non-symmetric mirror-curve. (b) The mirror-curve is made into an over-under knot

Figure 24.13 shows a stylized drawing of a turtle, made by the Tchokwe people from Angola. Its geometrical basis is the 3 × 3 grid. By introducing internal mirrors (how many of them do you need for

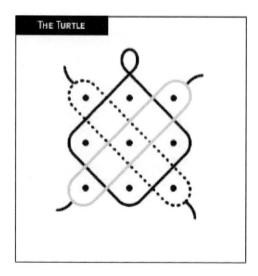

Fig. 24.13.   A turtle is created by a mirror-curve through a 3 × 3 grid

this?), try to transform this set of components into a monolinear curve.

In order to obtain interlacing knot ornaments (mirror curves similar to Celtic designs) you can use a few basic tiles (modules) as shown in Fig. 24.14a. These tiles represent all possible "states" of a single small square with regards to the portion of a mirror curve that it contains along with the relation "over-under". By (re)combining them, you can obtain all possible mirror curves (or knot mosaics) [Jab4, Rad4]. The same can be done by using only 5 modules ("Knot-Tiles") as shown in Fig. 24.14b. For the knot game made from "Knot-Tiles", please see http://www.mi.sanu.ac.rs/vismath/op/tiles/kt/kt. htm.

Consider the possibility to insert internal mirrors into a regular square grid lacking internal mirrors and covered by a single curve (with $c = \mathrm{GCD}(a, b) = 1$, i.e., when the sides of the grid are relatively prime numbers). We can try to plan the choice of positions for mirrors in advance, in order to obtain more interesting and visually pleasing single curves. In many cases, this can be achieved by using symmetry (translational = repetitional, rotational, or any other). After some experimenting, you can obtain single curves similar to those created

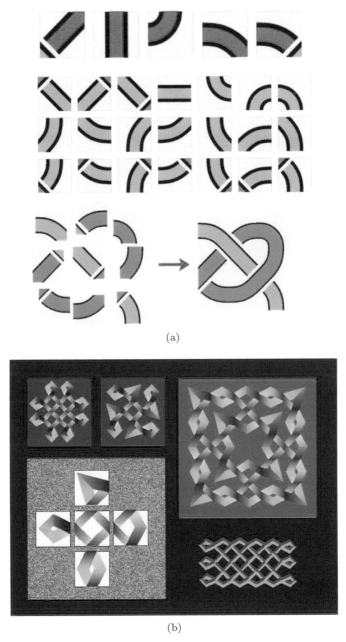

(a)

(b)

Fig. 24.14. (a) Slavik Jablan's "knot tiles." (b) Knot patterns formed with knot tiles by the Tchokwe people (Fig. 24.15) from 6 × 5 square grids (initially containing one component) and 5 × 5 grids (initially containing 5 components).

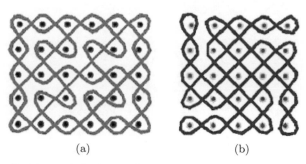

(a)   (b)

Fig. 24.15.   Positions of the internal mirrors used to create mono-linear curves through (a) a 6 × 5 grid, and (b) a 5 × 5 grid

We can discover the method of constructing monolinear curves in Fig. 24.15 by beginning with 6 × 5 and 5 × 5 grids, and determining the positions of internal (two-way) mirrors, placed within the grids in symmetrical ways, as shown on Fig. 24.16.

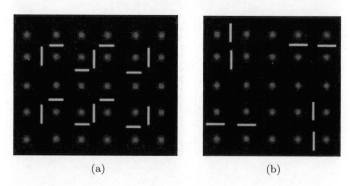

(a)   (b)

Fig. 24.16.   Arrangement of mirrors for the (a) 6 × 5 grid, and (b) 5 × 5 grid

Figure 24.17 shows the 5 × 4 grid initially containing a monolinear mirror curve. After a few steps, by introducing six internal mirrors in appropriate (symmetrical) positions, the Celtic people again obtained a monolinear mirror curve with symmetric design in a form of a Celtic knot.

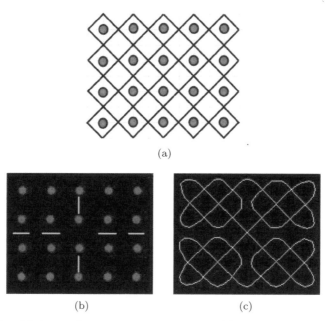

(a)

(b)   (c)

Fig. 24.17.   (a) A monolinear curve in a 5 × 4 grid; (b) A symmetric arrangement of internal mirrors; (c) A symmetric monolinear curve results

Figure 24.18 shows examples of Celtic knots from the cover of the book "Celtic Art — the Methods of Construction" (Dower, New York, 1973) by G. Bain. All Celtic knots can be (re)constructed by taking a grid (part of a plane tessellation; in the simplest case a part of a square grid), placing internal mirrors in it, and introducing the relation "over-under" (interlacing) at every crossing of the mirror curve.

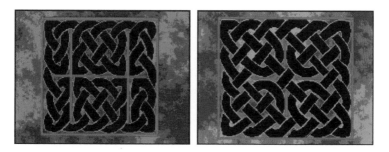

Fig. 24.18.   Examples of mirror-curves in the form of Celtic knots

The next knot (Fig. 24.19) is created in **Adobe Illustrator**. Beginning with a 4×3 grid, we introduce four internal mirrors and create the mirror curve, drawn by a 20pt line with the use of neon effects. Isn't that simple?

Fig. 24.19.   Celtic neon knots and their genesis

In Fig. 24.20, different mirror schemes are shown which can be used to create Celtic neon knots in the same manner as we constructed the knots in Fig. 24.19.

It is also possible to combine several rectangular schemes and extend our mirror-curve designs to obtain friezes or more complicated ornamental knots. By combining, joining, or overlapping parts of basic rectangular grids, we obtain composite mirror-curve designs [Jab4, Jab8, Rad2] (Sona drawings of animals made by the Tchokwe people, or monolinear Tamil threshold designs, called "pavitram" or "Brahma mudi") (Fig. 24.21).

The same approach can be used in order to obtain composite Celtic knots (e.g., in the form of a cross) (Fig. 24.22).

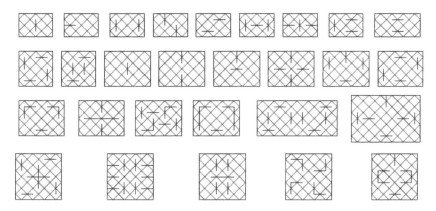

Fig. 24.20.   Different mirror schemes shown from which to create Celtic knots

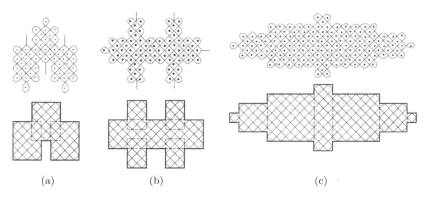

(a)   (b)   (c)

Fig. 24.21.   Sona drawings of animals made by Tchokwe people and Tamil threshold designs

Fig. 24.22.   Celtic knots in the form of a cross

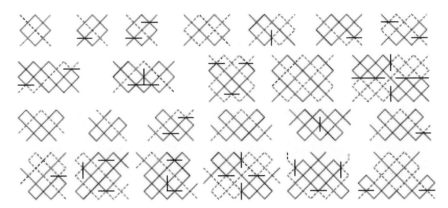

Fig. 24.23.   Tangles used in Celtic knotwork (tangles are knots with four loose ends)

Figure 24.23 shows the basic elements (mirror schemes) used in Celtic knotwork, that we shall call "Celtic tangles" (since each of them can be treated as a tangle with four loose ends and can be connected to others).

Problem: By using the method described in previous examples of rectangular grids and mirror curves, construct a knot of your choice. Convert the image to JPG format, write a short description in *MS Word*, and describe all the phases of your work. You can use any graphical computer program (*Inkscape, Adobe Illustrator, Corel Draw*). Some examples are shown in Fig. 24.24.

Constructions: Experiment by creating your own mirror curves and Celtic knots.

Fig. 24.24.   Additional examples, created in *Inkscape* and *Adobe Illustrator*

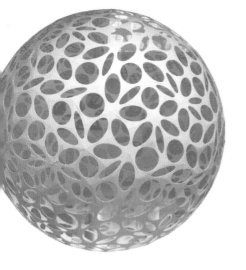

Chapter 25

Lunda Designs by Slavik Jablan and Ljiljana Radovic

After learning how to draw mirror curves, we consider designs referred to as *Lunda designs*, based on monolinear mirror curves. Every red dot in $RG[a, b]$ is the common vertex of four small squares (called "fields") surrounding it (see Fig. 24.1). Every mirror curve contains diagonals of these small squares. If we trace a monolinear mirror curve and follow the sequence of adjacent diagonals which it contains (called *steps*), after coloring small squares corresponding to the successive steps in the alternating manner (black-white-black-white...), we obtain a "black-white" design, named by P. Gerdes "Lunda designs" shown in Fig. 25.1 [Ger].

In order to draw a Lunda design, we first choose an arrangement of mirrors (Fig. 25.1a) that results in a monolinear mirror curve (Fig. 25.1b). The method of construction was described in the last chapter, "Mirror curves". Then we trace the mirror curve (Fig. 25.1c) and color the small field squares (Fig. 25.1d). In our drawing we started coloring from the second small square in the bottom row (the first square colored black), then followed the mirror curve alternating colors, black-white-black.... After closing the curve, we remove the red dots and obtain a black-white design much like Neolithic, black-white antisymmetrical patterns (Fig. 25.1e). Finally, we remove the grid (Fig. 25.1f). Additional Lunda designs can be found on the website.

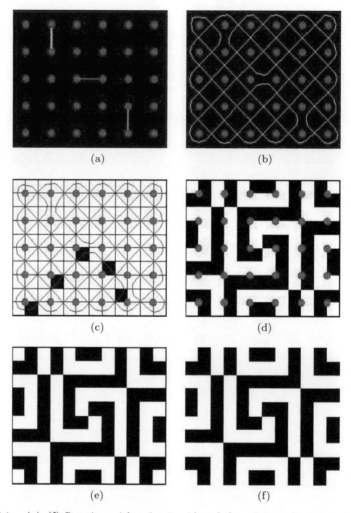

Fig. 25.1.   (a)–(f) Starting with a 6 × 5 grid and three internal mirrors steps are taken to create a black-white Lunda design

Figure 25.2 shows different arrangements of two-sided mirrors in the square grid 2 × 2, and their corresponding Lunda designs. Notice that the correspondence between mirror curves (i.e., mirror arrangements) and Lunda designs is many-to-one: different mirror curves can give the same Lunda design. Try to find the criterion in which two or more different (up to isometry) mirror arrangements

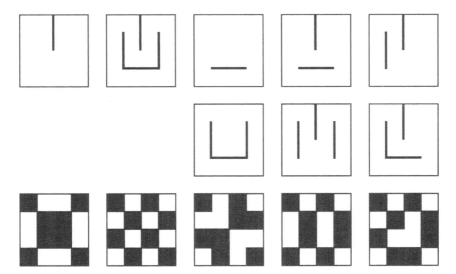

Fig. 25.2. 2 × 2 grids with internal mirrors and their resulting Lunda designs

result in the same Lunda pattern. If two mirror arrangements M_1 and M_2 gives the same Lunda design, find the transformation rules (moves of mirrors) converting M_1 to M_2.

In the 10×11 grid, place four mirrors as shown in Fig. 25.3; draw the mirror curve; follow the steps; and color small squares (fields) by two colors (yellow and light blue) in an alternating manner. The result is an antisymmetrical design which has vertical mirror reflection and horizontal antireflection (reflection in a horizontal mirror which changes colors).

Remark: A black and white design is called antisymmetric if the design is preserved under symmetry transformations which make black transform to white and white to black. For example, the Lunda design in Fig. 25.3 preserves the pattern under reflection in a horizontal mirror but changes blue to white and white to blue.

The same construction can be applied to an arbitrary mirror curve. If we count the steps along a mirror curve and numerate them by $1, 2, 3, 4 \ldots$, by taking all numbers modulo 2, we obtain the sequence $1, 0, 1, 0, \ldots$ which corresponds to the alternating coloring of steps "black", "white", "black", "white", $\ldots$ By taking all numbers in the sequence $1, 2, 3, 4, \ldots$ modulo 4, we obtain the sequence

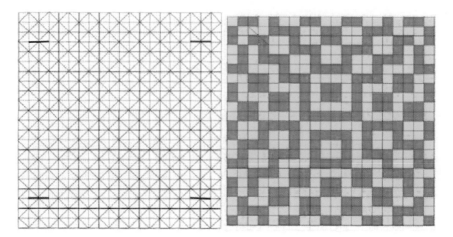

Fig. 25.3.   Four symmetrically placed mirrors in a 10 × 11 grid creates a symmetric Lunda design

$1, 2, 3, 0, 1, 2, 3, 0, \ldots$. After coloring all fields denoted by the same numbers, we obtain a 4-colored Lunda design. In the same way, we can work with $8, 16, \ldots$ colors. With two colors we obtain antisymmetric rosettes with one or two antisymmetry axes, and if we use more colors (i.e., 2^n colors, $n > 1$) we obtain so-called multiple antisymmetry.

On the border of $RG[a, b]$, every red dot is the common vertex for two adjacent small squares (fields), and in the interior of $RG[a, b]$ every square of dimensions 2×2 can be divided in two rectangles of dimensions 2×1 or 1×2, placed between two red dots, each of them containing one black and one white field. Therefore, this local arrangement results in the global arrangement: the number of black fields in any row or column is equal to the number of white fields (Fig. 25.4). Translated into the language of numbers, if we denote black fields by 1, and white squares by 0, and consider Lunda designs as matrices of ones and zeros, we obtain so-called Lunda matrices with the amazing property: sum of the numbers in every column is a, and the sum of numbers in every row is b. In particular, for $a = b$, we obtain square 0-1 matrices giving the same sum of numbers in each row and column.

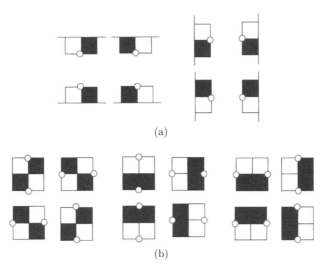

(a)

(b)

Fig. 25.4. (a) Possible border situations; (b) possible situations between vertical and horizontal neighboring grid points

Another possibility for creating visually interesting designs is the derivation of fractals from Lunda designs. As a result, we obtain so-called Lunda fractals (Fig. 25.5). If the same algorithm used for the creation of an initial Lunda design is applied again at a smaller scale, we obtain self-referential Lunda designs, i.e., Lunda fractals. Lunda fractals can be easily drawn in any drawing computer program. After making an initial Lunda design, we can scale it (in ratio 1:4), and then apply the same construction rule used for the creation of the initial design to the scaled parts.

Fig. 25.5. A Lunda fractal

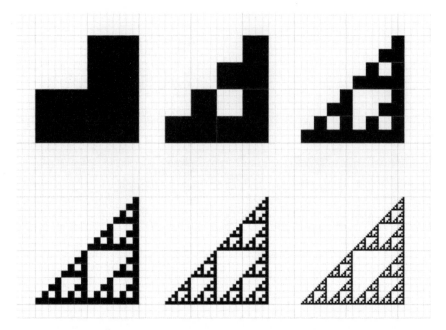

Fig. 25.6.   Evolution of Sierpinski's triangle

In *Inkscape* you can do a similar thing (Fig. 25.6): beginning from a black square of dimensions 2 × 2 with deleted upper left field (which is not a Lunda design, since the number of black squares in each row or column is not the same), you can scale it (in the ratio 1:2) and then organize the three scaled images, into the same pattern as the initial motif was made from three black squares. Continuing to apply the same algorithm, we obtain one of the most famous fractals called the Sierpinski triangle. Fractals were described in Chaps. 13 and 14.

You can reconsider modular games "OpTiles", "OrnTiles" and "KnotTiles" created by S. Jablan (http://www.mi.sanu.ac.rs/vismath/op/tiles/index.html), try and to create your own sets of modular elements.

Inspired by potato prints by M.C. Escher and their basic tile called the "Escher tile" resulted in Fig. 25.7. "OrnTiles" is a modular game created from only one basic element (tile) and its transforms are obtained by rotations and reflections in the vertical and

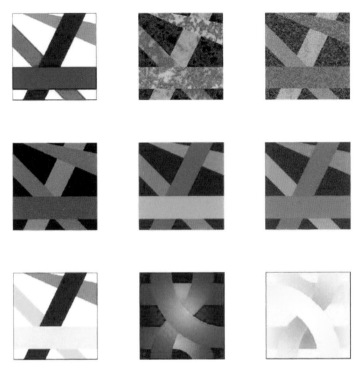

Fig. 25.7.   Examples of an Orn tile designed by Slavik Jablan

horizontal mirror axis (Fig. 25.8). Using them, we may derive a large collection of different ornaments. As a result, every "OrnTile" fits perfectly (edge-to-edge) with all of its neighbors, and independent of the positions of the tiles, which always gives an interlacing pattern (periodic or not), placed in a rectangular grid. This is made possible by the appropriate choice of an initial tile which acts like a tangle with two loose ends at each side, permitting different topological variations (use curved lines and arcs for the borders of colored regions instead of straight lines). Try to make similar tiles with more than two loose ends at each side. Can you make them from polygons other than squares? Which polygons can you use? Can you compose interlacing patterns beginning with different polygonal basic tiles?

In order to make your first "OrnTile", take a square of dimensions 5×5 and join the edges by "strips" (not necessarily rectilinear), as shown on the left side of the Fig. 25.9. Then join two opposite

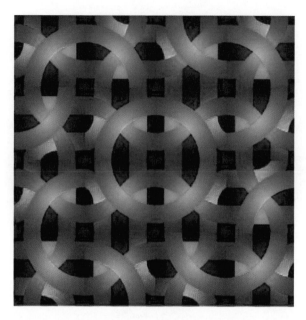

Fig. 25.8.   Example of an Orn tile designed by Slavik Jablan

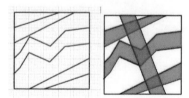

Fig. 25.9.   An Orn tile made from a square of dimension 5 × 5. The tile is continuous through left-right and up-down

edges, top and the bottom, by the strip crossing the others. If you like, you can make your strips alternating (going "over-under"), or curvilinear. Figure 25.10a shows another Orn tile. Use it to construct your own OrnTile design. At the end of the construction you can make the edges of the basic square invisible (Fig. 25.10b). A similar construction is shown at Fig 25.11. Instead of coloring the strips, you can use different shapes and textures (e.g., marble, recycle, *etc.*) from the texture library (e.g., in *Corel Draw* by the command Fill Tool > Texture Fill Dialog). Tiles constructed in this way always fit

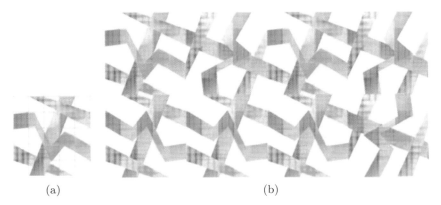

(a)   (b)

Fig. 25.10.   (a) A single Orn tile, (b) The tile continues connecting left-right and up-down through a square lattice

Fig. 25.11.   A sequence of Orn tiles creating a pattern in a square lattice

perfectly with their neighbors as it transforms from the initial tile (obtained in *Corel Draw* by rotating or adding mirrors to the initial tile by the commands: Transform > Rotate or Transform > Scale and Mirror). What will happen if you shift your tiles along the vertical or horizontal line?

Construction: Lunda fractal. Construct a Lunda design of your choice, and then apply self-similarity by copying and scaling the initial tile, and then arranging the resulting tiles following the same rule according to which the initial Lunda design is made. You can work in any graphical computer program (*Inkscape, Adobe Illustrator, Corel Draw*).

As an alternative construction you can construct "OpTiles" and make your own designs from them. In this case, interesting patterns will be obtained if you shift your tiles along the vertical and horizontal line (i.e., you obtain patterns that are not trivial "edge-to-edge" patterns).

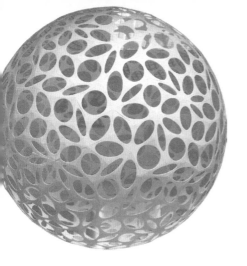

Chapter 26

Visual Illusions by Slavik Jablan and Ljiljana Radovic

26.1. Impossible Objects

The perception of a 3-dimensional object represented as a 2-dimensional projection is not always easy. If we try to explain what we see as an object, we would conclude that we make a choice between an infinite series of real 3-dimensional objects having the same flat retinal projection. Our perception generally selects only one "natural" (most usual or, simplest) interpretation. In the case of isolated objects, when a common reference system does not exist, sometimes there is ambiguity-impossibility of determining a unique "natural" interpretation. The situation is even more confusing if such a "natural" interpretation is an impossible object — a figure that contradicts our sense of visual 3-dimensional perception [Jab9], [Bar].

Some of the well-known impossible objects are Thiery figures (proposed at the end of XIX century), Penrose tribar — objects created by Oscar Reutesvard in 1934, Vasarely constructions, etc. All of these can be derived as modular structures from a Koffka cube. But what is a Koffka cube? Crystallographer Louis Albert Necker, looking at transparent crystals from a cube, discovered that a single transparent crystal produces two different images: one convex and the other concave. This visual illusion is referred to as Necker's illusion.

In the theory of visual perception, it is referred to as a Koffka cube (after the Gestalt psychologist Kurt Koffka). The Koffka cube can be viewed as a simplification of Necker's illusion — the image of a cube, when we see a regular hexagon can be concave or convex. The Koffka cube is multi-ambiguous — it can be interpreted as three rhombuses with a joint vertex, as a convex or concave trihedron, or as a cube. If we accept its "natural" 3-dimensional interpretation — a cube, then for a viewer there are three possible positions in space: upper, lower left and lower right, having equal right to be a valid interpretation.

In this section we construct a Koffka cube which is the basic element for the construction of a variety of impossible figures.

Let's start from a regular hexagon ⬡. Divide it in three rhombuses with common vertex at the center of the hexagon ⬡. Color the rhombuses, delete the hexagon and remove the edges of the rhombuses. From this grouping of three rhombuses we can create the Koffka cube (Fig. 26.1.)

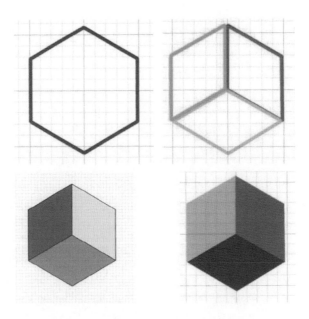

Fig. 26.1.   From hexagon to Koffka cube

The resulting building block, a modular element — the Koffka cube, can be multiplied and combined with other Koffka cubes. First, combine several Koffka cubes without overlapping by using rotations and reflections. Even at this point, many strange objects can be constructed that are difficult to recognize as being convex or concave, thus offering ambiguous interpretations (Fig. 26.2).

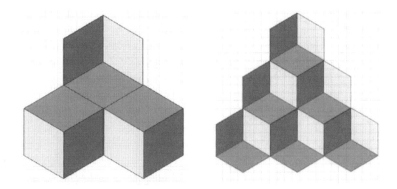

Fig. 26.2.   Combination of several Koffka cube in different positions

Koffka cubes can be combined with overlapping if we introduce the relation: "in front"-"behind". Depending on the program you use, there are different options. Usually, you can use the commands **Bring to <front, end, bottom, top>** which you can apply to ungrouped objects (**Ungroup**) or to whole objects.

At the ***Inkscape*** *menu*, one can find options to bring a selected object to the desired position ⬇⬆➡⬆.

For example, if in the following composition we select the square closest to us and use the first option *Lower selection to bottom* (End) we obtain the image shown in Fig. 26.3a), but If we use the last option *Raise selection to top* (Home) we obtain the image in Fig. 26.3b.

By using these spatial relations and combining Koffka cubes, we obtain different impossible objects, such as the Penrose tribar or Escher's *infinite* stairway and waterfall. You can use the Kofka cube as well as the rhombuses in it to construct these objects as shown in following figures showing : a Penrose tribar, Waterfall, Infinite stairways, Double stairs, and the letter H.

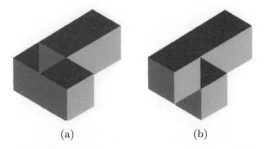

(a) (b)

Fig. 26.3. (a), (b) Two arrangements of Kofka cubes

Penrose tribar

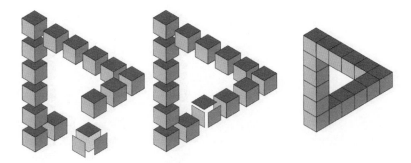

Waterfall

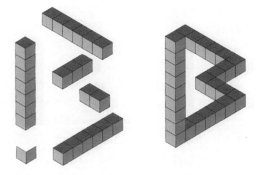

Infinite stairways

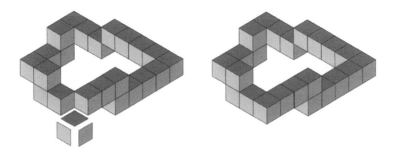

Double stairs (convex-concave)

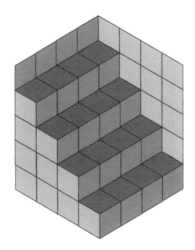

Letter H

Impossible cube

If you play with the in front of and behind relation, you can draw the impossible cube shown in Fig. 26.4.

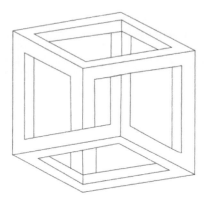

Fig. 26.4.   An impossible cube

Exercise: Using Koffka cubes, make your own impossible figure. You can use any graphical software.

26.2. Op-art and Visual Illusions

The theme of this section is Op-art (optical art) and visual illusions used in art [Ern], [Jab6]. We will introduce exercises, using some ideas of Ajashi Kitaoka and other masters of visual illusion, and you can try, using this experience to conduct research into creating new optical compositions.

Many optical visual illusions are based on the specific choice of basic elements, their repetition, and choice of (mainly complementary) colors. This option is not precisely defined in Inkscape so you will need to experiment with color. In Corel Draw or Adobe Illustrator every color is given by a code (RGB, CMYK), and then you can directly obtain its complementary color in the following way:

Fig. 26.5.   Set RGB code

Start with very simple bands producing the so-called *Kindergarten illusion*, where straight lines are disturbed and move from their vertical positions (see Fig. 26.6). First draw a rectangle and color it by a chosen color. Then glue, on its side, another rectangle of the same dimensions but colored by complementary colors. Repeat this construction until a band consisting of rectangles alternating in colors is obtained. From this band create a group of bands by duplicating it. Glue the original and the duplicates, shift vertically up or down by one third of the length of the edge of its initial small rectangle, and arrange the double bands by shifting them vertically. You can see that the illusion is stronger if all rectangles are framed by thin lines.

The following examples show this construction and its variants in which the whole structure is placed in the background, having, the same color as one of the small rectangles. Then the border lines (frames) of the small rectangles are removed (Fig. 26.7a). Although we have constructed parallel brands, they do not appear to be so in the final image, especially if we rotate them horizontally (Fig. 26.7b).

Fig. 26.6.   Kindergarten illusions

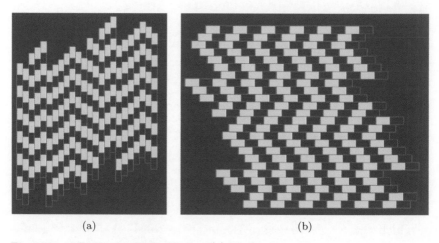

(a)   (b)

Fig. 26.7.   The Kindergarten illusion, (a) Kindergarten illusion in a vertical position, (b) Kindergarten illusion in the horizontal position

In a similar way, but only in black and white, we can produce different "figure"–"ground" effects or deformations from straight lines. We constructed the following Lunda design in Chap. 25 (Fig. 26.8):

Fig. 26.8. Figure-ground effects derived from Lunda designs. Apply the kinetic illusions of Fig. 26.9 and Fig. 26.10 to try distorting the Lunda design

Consider black and white squares in which you place small circles of complementary color and emphasize one side: left or right, or upper or lower, i.e., desymmetrize the initial squares (Fig. 26.9). Even from a simple checker-board pattern, as in Fig. 26.10, you can obtain very interesting visual illusions. Try to create similar visual illusions from Lunda designs, as shown in Fig. 26.10, while making use of (complementary) colors.

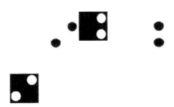

Fig. 26.9. Black and white circles placed in Black and white squares to desymmetrize the initial squares

Fig. 26.10.   Checkerboard pattern creating the illusion of motion based

Let's see how we can create the illusion of motion in static art-works and produce "kinetic" visual illusions in the style of Ajashi Kitaoka.

We start from a "coffee-bean" in an elliptic form, which we obtain by the deformation of a circle. Its border needs to be divided into two semi-ellipses with different colors (dark and light). This can be done even before deforming the circle, by drawing arcs corresponding to central angles of 180 deg. You can join these colored arcs with a circle (i.e., make a group), and then deform the whole group and obtain the ellipse. You can then put this ellipse in a rectangle and color it similarly to the background. This rectangle serves as the frame, enabling repetition of the ellipses in later compositions.

In order to obtain a dynamical structure that suggests motion, we need to break the symmetry of our ellipses, and put some "weights" on their sides (i.e., colored edges). The first horizontal line of the following pattern is obtained by multiplying the ellipse and rotating it in every step by 45 deg. (or by 30 deg.). The next line is obtained by copying the first and shifting it one step to the right. In the first case we consider the ellipse rotated by −45 deg. (or, better, by −30 deg.), and then eliminate the last element in the sequence.

We repeat this algorithm and obtain the following images in Fig. 26.11:

Fig. 26.11. Starting from a coffee-bean in elliptic form one obtains a dynamical structure suggesting motion

In all such cases it is necessary to experiment with complementary colors, size of the initial elements and of the complete image. Try different combinations, and they will result in dynamic, as well as static images. Learn from the work of other artists that produced such images, e.g., A. Kitaoka, as well as from your own experience.

We end with an example involving linear perspective. Thanks to linear perspective, distant objects appear smaller, and their size decreases proportionally with the distance from the viewer. Hence, two identical objects, or objects of the same size, placed on different levels, in an environment organized by linear perspective with strongly converging perspective lines, look completely different. Hence, the object in front looks much smaller than the same object at the back. In Fig. 26.12 the two horizontal lines are of the same length, but the one in front looks smaller than the one at the back. There are many other examples of this kind of illusion.

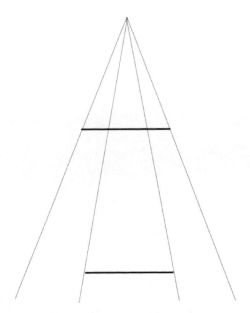

Fig. 26.12.   Illusions based on linear perspective. The object in front appears
smaller than the same object in the back

Exercise: Construct your own example of a visual illusion.

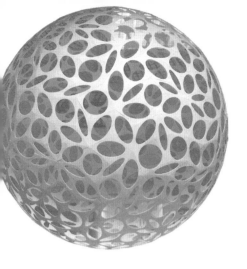

Epilogue

We have come to the end of this book which has endeavored to present mathematics in the "mind's eye," a presentation of visual mathematics. Before his untimely death, my colleague Slavik Jablan made it his mission to promote the idea of visual mathematics. For almost 20 years, he edited the Journal of Visual Mathematics, taught a course in Mathematics of Design, and edited a book entitled Symmetry, *Ornament and Modularity* [Jab3] which brought the visual aspect of mathematics into clear view. His student, Ljiljana Radovic and myself have, in our own ways, worked to further these ideas. It is fitting that we end this exposition with a survey of mathematical illusions.

Although mathematics can be used in an applied sense, this is the utilitarian aspect of the subject. But visual math is a mixture of fact and fancy, a combination of beauty and elegance tied together with the rigor mathematics is famous for. We have tried to present mathematics in terms of its aesthetics much as one would approach music. Another book comes to mind, in addition to the book cited above, *Aesthetics of Interdisciplinarity: Art and Mathematics* edited by Kristof Fenyvesi and Tuuli Lahdesmaki [Fen2] with a forward by me. These two books contain much of the ground covered in this volume but in their own ways. Additional papers by my colleagues

Jablan [Jab3,6,9] and Radovic [Rad1,2] can also be found in the bibliography.

Finally, this book is dedicated to the work of the many mathematicians, artists, architects, designers, quilters, recreational mathematicians and all those who have found ways to bring mathematics to life through the geometrical foundations of design.

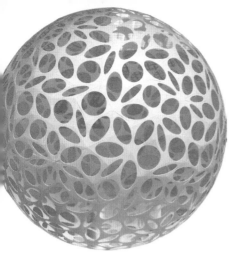

Bibliography

[Ada]   Adams, C. The Knot Book. New York, W.H. Freeman (1994)

[Arn]   Arnol'd, V.I. The branched covering $CP^2 \rightarrow S^4$, hyperbolicity and projective topology., Sibirsk. Mat. Zh. 29, 5, 36–47, 237 (1988).

[Bar1] Barrett, C. Op art, Studio Vista, London (1970).

[Ben]   Bender, S. Plain and Simple, New York: Harper and Row (1989Blo).

[Bev]   Bevan, P. Open Meanders., http://demonstratons.wolfram.com/ OpenMeanders (2013).

[Blo]   Blochet, E. The peintures des manuscrits orientaux de la Biblioteque Nationale (1912).

[Bru]   Brunes, T. Secrets of Ancient Geometry, Copenhagen: Rhodos (1967).

[Cho]   Chorbachi, W. In the Tower of Babel: Beyond Symmetry in Islamic Design. Comp. Math Applic. Vol. 17, No. 4–6, pp. 751–789 (1989).

[Con]   Conway, J. An enumeration of knots and links and some of their related properties. In Computational Problems in Abstract Algebra, Proc. Conf. Oxford 1967 (Ed. J. Leech), 329–358, Pergamon Press, New York (1970).

[Cox1] Coxeter, H.S.M. Golden Mean, Phyllotaxis, and Wythoff's Game, Scripta Mathematica, Vol. XIX No. 2, 3 (1955).

[Dew]   Dewar, R.E. Islamic giirih tiles in their own right as a history lesson and design exercise in the classroom, Vol. 20, Numbers 1–4, pages 201–216 (2009).

[DiF]   Di Francesco, P. *Folding and coloring problems in mathematical physics.* Bull. Amer. Soc. (2000).

[Ech1]   Margit Echols1 Personal communication.

[Ech2]   Margit Echols. Personal communication.

[Ech3]   Piecing Symmetrical Prints. Threads No. 55, pp. 56–59.

[Edw]   Edwards, E.B. Pattern and Design with Dynamic Symmetry New York: Dover (1967).

[Ein]   Einstein, A. The cause and formation of meanders in the courses of rivers and the so-called Baer's Law, Die Naturwissenshaften, Vol. 14 (1926).

[Fen]   Jablan, S., Radovic, L. and Fenyvassen, K. Following the Footsteps of Daedalus; Labyrinth Studies Meets Visual, Proceedings of the Bridges: 2013 Conference Mathematics, Music, Art and Architecture (2013).

[Flo]   Flower of Life https://www.ancient-origins.net/human-origins-religions/ancient-secrets-lie-within-flower-life-001078.

[Fra]   Frantz, M. and Crannell, A. Viewpoints: Mathematical Perspective and Fractals in Art (2003).

[Fram]   Frame, M.L. and Mandelbrot, B.B. Fractals, Graphics, and Mathematics Education, MAA Notes.

[Fri]   Friedman, N. Private Communication (2003).

[Gar1]   Gardner, M. Aha, Insight Scientific American (1978).

[Gar2]   Gardner, M. Topological Diversion, Including a Bottle with No Inside or Outside, In Mathematics": An Introduction to its Spirit and Use, Scientific Am, Inc. San Francisco: W.H. Freeman and Co. (1979).

[Gar3]   Gardner, M. Curves of Constant Width, One of which Makes It possible to Drill Square Holes, Scientific Am., Inc. San Francisco: W.H. Freeman Press (Feb. 1963).

[Gar4]   Gardner, M., Extraordinary Nonperiodic Tiling. *Sci. Am.*, pp. 110–121 (1978).

[Gar5]   Gardner, M., A Periodic Tesselations from the book, Penrose Tiles to Trapdoor Ciphers. MAA (1988).

[Ger]   Gerdes, P. Lunda Geometry, Ethnomathematics Research Project, Mozambique: Universidade Pedagogica (1996).

[Gil]   Gilbert, W. J. Simple Tilings with Lattice Symmetry, Structural Topology No. 8 (1983).

[Ham]   Hambridge, J. The Fundamental Principles of Dynamic Symmetry. New York: Dover (1967).

[Jab1]   Jablan, S. and Radovic, L. *Meanders, Knots, and Links.* arXiv:1302.1472vI [math.GT]. (2013).

[Jab2]   Jablan, S.V. and Sazdanovic, LinKnot–Knot Theory by Computer. World Scientific, New Jersey, London, Singapore, http://math.ict.edu.rs/ (2007).

[Jab3]   Jablan, S.V. *Symmetry, Ornament, and Modularity*, World Scientific, Singapore (2002).

[Jab4]   Jablan S.V. *Modularity in Science and Art*, Visual Mathematics, 4, 1, (2002).

[Jab5]   Jablan, S. and Radović, L. *Do you like paleolithic op-art?* Kybernetes, Vol. 40, No. 7/8, pp. 1045–1054 (2011).

[Jab6]   Jablan, S. and Radović, L. *The Vasarely Playhouse: Invitation to a Mathematical and Combinatorial Visual Game.* In: Fenyvesi K., Lähdesmäki T. (eds.) Aesthetics of Interdisciplinarity: Art and Mathematics. Birkhäuser (2017).

[Jab7]   Jablan, S. and Radovic L.J. *Patterns, Symmetry, Modularity and Tile Games.* In Adventures on paper, Math–Art activities for experience — centered education of mathematics, Edited by Tempus EU Team, Eszterházy Károly College, Eger, Hungary (2014).

[Jab8]   Jablan, S., Radovic, L.j., Sazdanovic, R. and Zekovic, A. *Knots in Art*, Symmetry-Basel, MDPI, 4, 2, pp. 302–328 (2012).

[Kap1]   Kappraff, J.M. A participatory Approach to Modern Geometry. Singapore: World Scientific Publ. Co. (2015).

[Kap2]   Kappraff, J.M. Beyond Measure, Singapore: World Scientific (2000).

[Kap3]   Kappraff, J. Connections: The geometric bridge between art and science, first published in 1990 by McGraw Hill, Singapore: World Scientific (2000).

[Kap4]   Kappraff, J. "The Arithmetic of Nicomachus of Gerasa and its Applications to Systems of Proportion" Nexus Network Journal Vol. 4, No. 3 October 2000. Kapu Persenol communication.

[Kap5]   Kappraff, J. and McClain, E.G. "The Proportions of the Parthenon: A work of musically inspired architecture" International Journal for Music Iconography, Vol. XXX, No. 1–2 Spring-Fall (2005).

[Kap6]   Kappraff, J. Silver Mean, Bridges Conference (2011).

[Kap7]   Kappraff, J. and Adamson, G. Generalized Genomic Matrices, Silver Means and Pythagorean Triples, Forma Vol. 24, pp. 41–48 (2009).

[Kap8]   Kappraff, J. and Petoukov. Symmetries, generalized numbers, and harmonic laws in matrix genetics. Symmetry: Culture and Science ed. G. Darvas. Vol. 20, No. 1–4, pp. 23–49 (2009).

[Kap9]   Kappraff, J., Radovic, L.J. and Jablan, S. *Meanders, knots, labyrinths and mazes*, J. Knot Theory and Ramifications, 25(9) (2016).

[Kap10]   Kappraff, J., Adamson, S. and Sazdanovich, G. Golden Fields. FORMA Special issue on the golden mean. Vol. 19, No. 4, pp. 367–387 (2004).

[Kap11]   Kappraff, J. Ancient armonic Law as a Generator of Cultural Metaphors (2013).

[Kap12]   Kappraff, J. Systems of Proportion in Design and Architecture and their Relationship to Dynamical Systems Theory, Bridges Conference, Mantua, Italy (1999)

[Kap13]   Kappraff, J. Golden Fields, Generalized Fibonacci Sequences, and Chaotic Matrices, Forma Vol. 19 No. 4 pp. 367–387 (2004).

[Ker1]   Kerenyi, K. *Labyrinth-Studien. Labyrinthosals Linienreflex einer mythoologischen Idee*, In: Humanistische Seelenforschung, Langen Muller, Munchen, Wien, 226–273 (1969).

[Ker2]   Kerenyi, V. Gedanken ujber den grieschiscen tanz, In: Humanistische Seelunforschung, Langen Muller, Munchen, Wien, 274–288. Labyrinth K (1966).

[Ker3]   Kerenyi, K. *Dionysos: Archtypal Image of Indestructible Life*, Ralph Manheim trans., Princeton University Press. (1976).

[Fen1]   Fenyvesi, K. Jablan, S. and Radovic, L.J. *Myth meets mathematics: tiles, meanders and labyrinths*, Symmetry: Art and Science, Special Issue for the Congress Festival of ISIS Symmetry, Editors: Ioannis M. Vandoulakis and Dénes Nagy, Symmetry: Art and Science, 112–115, (2013).

[Fen2]   Fenyvesi, K., Jablan, S. and Radovic, L.J. In the Footsteps of Daedalus: Labyrinth Studies Meets Visual Mathematics, Proceedings of Bridges 2013 World Conference, Editors: G. Hart & R. Sarhangi, 361–368, (2013).

[LaC]   La Croix, M. *Approaches to the Enumerative theory of Meanders*, http://www.math.uwaterloo.ca/malacroi/   Latex/Meanders.pdf (1904).

[Lal1]   Lalvani, H. Non-periodic Space Structures, Int. Jour. Of Space Structures, Vol. 2, No. 2, https://doi.org/10.1177/026635118700200204.

[Lal2]   Lalvani, H. Continuous Transformations of Non-periodic Tilings and Space-fillings, Preprint (1990), published in: Istvan Hargittai: Five-fold Symmetry, World Scientific (1991).

[Lal3]   In Kap3, 2nd ed. Figs. 6.41 and 6.42.

[Lal4]   Lalvani, H. Morphological Universe: Genetics and Epi-genetics in form-making, Symmetry: Culture and Science, Vol. 29, No. 1, p. 124 (2018).

[Lal5]   Lalvani, H. In: Kappraff, J. *Connections*, 2nd Edition, World Scientific, 2001, Supplement S.5, New Morphological Discoveries of Haresh Lalvani, pp. 282–6.

[Lal6]   *Space Structures with Non-Periodic Subdivisions of Polygonal Faces*, U.S. Patent 5, 524, 396, June 11, 1996.

[Lal7]   Haresh, L. Morphological Universe: Genetics and Epigenetics in Form-Making, *Symmetry: Culture and Science*, Vol. 29, No. 1, pp. 138–48 and p. 162 (2018).

[LeC]   LeCorbusier, Modulor, Cambridge: MIT Press (1962). Modulor 2 Cambridge: MIT Press (1962).

[Liv]   Livio, M. The Golden Ratio: The Story of Phi, the World's Most Astonishing Number. New York: Broadway Books (2002).

[Loe1]   Loeb, A.L. Algorithms, Structure, and Models. Ub Hyoergraohics, AAAS Selected Series, pp. 49–68 (1978).

[Loe2]   Loeb, A.L. *Color and Symmetry.* New York: Wiley (1971).

[Mar]   March, L. and Steadman, P. Geometry of the Environment, Cambridge: MIT Press (1974).

[Mat]   Matthews, W.H. *Mazes and Labyrinths*, Longmans, Green, and Col, London. (1922).

[McC]   McClain, E.G. Pythagorean Plato, York Beach, ME: Nicolas-Hays (1978, 1984).

[Ols]   Olsen, S. The Golden Section. New York: Walker and Com. (2013).

[Phi]   Phillips, T. *Through Mazes to Mathematics.* http://www.math. sunysb.edu/tony/mazes/index.html (2013).

[Pic]   Pickover, C. *The Mobius Strip.* New York: Thunder's Mouth Press (2006).

[Rot]   Rolfsen, D Knots and Links. Publish & Perish Inc., Berkeley, 1976; American Mathematics Society, AMS Chelsea Publishing (1976).

[Rad1]   Radovic, L.J. and Jablan, S. *Antisymmetry and Modularity in Ornamental Art*, Reza Sarhangi & Slavik Jablan (eds.), Bridges 2001. Conference proceeding of Bridges: Mathematical Connections in Art, Music, and Science, 55–66, (2001).

[Rad2]   Radovic, L. and Jablan, S. *Visual communication through visual mathematics*, Filomat, 23(2), 56–67 (2009).

[Rad3]   Radovic, L.J. and Jablan, S. *Meander Knots and Links*, Filomat 29(10), 2381–2392 (2015).

[Rad4]   Radovic, L.J. and Slavik, J. *Mirror-curve Codes for Knots and Links*, Publications de l'Institut Mathématique (Beograd), 94 (108), pp. 181–186, (2013).

[Sar]   Sarhangi, R., Jablan, S. and Sazdanovic, R. Modularity in Medieval Persian Mosaics: Textual, Empirical, and Theoretical Considerations, Visual Mathematics, 7, 1, 2005 (reprint from "Bridges" Proceedings, 2004, pp. 281–292).

[Spe-B]   Spencer-Brown, G.

[Sta]   Stachov.

[The]   Theon of Smyrna. *The Mathematics Useful for Understanding Plato* (trs. From the Greek/French ed. Of Dupuis by R. and D. Lawlor), San Diego: Wizard's Bookshelf. (1979).

[Tou1]   Toussaint, G.T. A mathematical analysis of African, Brazilian and Cuban clave rhythms. Proceedings of the BRIDGES conference. Towson, MD, pp. 198–212 (2002).

[Tou2]   Toussaint, G.T. Classification and phylogenetic analsys of African ternery rhythmictTime lines. Proceedings of the BRIDGES Conference, Granada, Spain, pp. 25–36 (2003).

[Wat]   Watts, D.J. and Watts, C. A Roman Apartment Complex. Sci. Am., Vol. 255, No. 6, 132–140, Dec. (1986).

[Wil]   Williams, R. *The Geometrical Foundation of Natural Structure*, New York: Dover (1970).

[Wilf]   Wilfred, J. N. In Medieval Architecture, sign of Advanced Math. http://www.nytimes.com/2007/02/27science/27math.html (2007).

[Wit]   Witcower, R. *Architectural Principles in the Age of Humanism*, New York: John Wiley (1998).

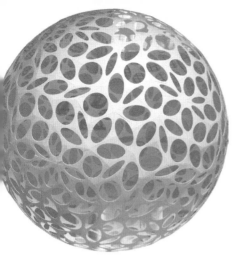

Attributions

The material that you find in this book comes about from many sources. Although I am responsible for the flow of ideas, I could not have written this book without the inputs of artists, designers, design scientists, and people who just like to build things. I wish here to give thanks and express my great respect to the many people whose work you will find represented.

Chapter 1 presents information about triangle-circle and square-circle designs from a workshop that I attended by Margit Echols.

Chapter 2 came about due to my friendship with Margit Echols, the author of more than 50 books on quiltmaking, now deceased. However, Margit's quilting differs from the usual quilt-maker by how she based them on fundamental mathematical principles. We will study these principles in greater depth as we move through this book. In this chapter I will focus on her Norman Conquest quilt pieced together from what she called, "Magic Squares." In this chapter, we focus on the mathematical fundamentals and present her quiltmaking techniques on the website.

Chapter 3 presented four geometrical proofs of the Pythagorean theorem including one by Annairizi of Arabia (900 AD) presented to me by the design scientist, Reza Sarhangi. A second proof came about from the article by Martin Gardner in his book,

Aha, Insight, [Gar1]. This article became the basis of a set of designs discovered in ancient documents by W.K. Chorbachi and written about in her classic 1989 article, *In the Tower of Babel* [Cho].

Chapter 4 made use of materials brought to my attention by Lionel March and Philip Steadman in their book The Geometry of Environment. They called my attention to the connection between non-congruent tilings and circuit diagrams.

In **Chapter 5**, William J. Gilbert introduced me to how to simply construct lattice tilings in 2- and 3-dimensions.

In **Chapter 6**, through the writings and correspondence with the Danish engineer, Tons Brunes, I became aware of his wonderful star based on 3,4,5-triangles. This star was featured by two chapters in my book, Beyond Measure.

I owe the content of **Chapter 7** to Slavik Jablan for his capturing of the history of design going back to Paleolithic times. I have also presented examples of the OpArt of R. Neal in "Square of Three," Frank Stella in "Hyena Stomp," and V. Vasarely in variation on the theme of OpTiles.

I was not able to locate the work of V. Vasarely or R. Neal. However, Hyena Stomp is in the Tate Liverpool Museum.

In **Chapter 8**, my colleagues Slavik Jablan and Ljiljana Radovic brought the use of Versatiles, Truchet tiles, Kufic tiles, and Optiles from Paleolithic times to the present so that my students, such as Kevin Miranda, are able to use them to make designs and create their own OpArt.

In **Chapter 9** you can see how Ljiljana Radovic and Slavik Jablan used meanders to make labyrinths and knots, how Kristof Fenyvesi traced the history of labyrinths and how I used meanders to construct labyrinths and mazes.

Chapter 10 introduces the reader to a series of ideas from topology that attempts to take the reader beyond their imaginations. This was something that Martin Gardner did very well, and you will see some of his constructions. One is a paper folding creation of a Klein bottle that is found in [Gar2]. Another is a novel proof of the Pythagorean theorem using cake cuts [Gar1], and another is information that I used in Chapter 12 on Curves of Constant Width [Gar3], and yet another is his discussion in 1978 on Roger Penrose's discovery about non-periodic tilings using kites and darts [Gar4,5]. I also thank C.Pickover for the clever image you will see in his book, The Mobius Strip [Pic]. You will also find three Mobius strip sculptures and information on how to construct a Klein Bottle. Unfortunately, I have never found the authorship of these sculptures.

Chapter 11 challenges the reader to construct a Szilassi polyhedron and a Csaszar Polyhderon, its dual.

In **Chapter 12**, Martin Gardner again shows his ability to stir the imagination by creating wheels that are not round but still roll. This information can be found in Mathematics: An Introduction to its Sprit and Meaning [Gar2]. Dick Esterley, an architect and toy inventor then added his discovery of a 3-dimensional sculpture based on an octahedron that also rolls.

Chapter 13 shows the simplicity behind fractals and how they are formed. I have chosen two sources to help me to describe this process, Viewpoints, by Franz and Crannell and Frame and Mandelbrot in Fractals, Graphics, and Mathematics Education. Some of the ideas from this chapter and the next were gotten by the attendance of a Viewpoints NSF workshop.

In **Chapter 14**, Crannell and Franz introduced me to the Iterative Function System (IFS). This chapter was based on a workshop to construct a fractal wallhanging at the Bridges Conference.

In **Chapter 15**, I drew upon a pair of perpendicular square root 2 rectangles by E.E.. Edwards [Edw] in his book, Pattern and Design in Dynamic Symmetry, Martin Gardner's four turtle problem, and the Baravelle spiral. Edwards also led me to the construction of a logarithmic spiral which forms the basis for much of this chapter.

Chapter 16 is devoted to the golden mean, a number and concept that I never tire of. In this chapter I draw on the work of Janusz Kapusta, a cartoonist, designer of opera sets and an amateur geometer. His amazing series of golden mean designs are on exhibit in this chapter.

Chapter 17 is about a fascinating game known as Wythoff's Nim. The great 20th century geometer, H.S.M. Coxeter, did a thorough analysis of this game which brought it to my attention.

Chapter 18 is about the Modulor of Le Corbusier. Le Corbusier always had a set of Modulor tiles in a draw of his office during World War 2, and he would try to understand their many relationships by playing with them. I had my students create tilings with the Modulor set and made good use of the exercises in the two volume set of books, Modulor.

Chapter 19 introduces us again to Martin Gardner who began a small revolution by bringing Penrose tiling by kites and darts to light in the late 1970s I attended a talk by Robert Dewar at the Symmetry Festival 2014 who introduced me to Girih tilings which enabled one to create non-periodic tilings in medieval times. Finally, I discovered how easy it is to recreate the Girih tilings using a computer program called Google SketchUp.

Chapter 20 brought to the surface, the ancient architecture of the Roman Empire, a system of proportions recognized by Theon of Smyrna, a Roman mathematician of the 2nd century AD. My first contact with such a system was by way of Donald and Carole Watts, two historians of architecture. In this chapter, I made use of the Dynamic Symmetry of Jay Hambridge.

Chapter 21 chronicles my great surprise to find that the Modulor of Le Corbusier, the Roman system of proportions, and the Pythagorean music scale all share a common foundation. This led me to consider the mathematics behind the Pythagorean musical scale. And for this information I am indebted to the Ethnomusicologist, Ernest McClain.

Chapter 22 is a lead into the material covered in the next chapter and so there are no outside sources that have influenced this chapter.

Chapter 23 describes a trip that I took with my family to Amish country in Lancaster Pennsylvania and how it led to some advanced concepts in geometry. I am thankful to Haresh Lalvani, a design scientist and professor of architecture at Pratt University for showing me the relationship between zonogons and polyhedra. Prof. Lalvani is also a sculptor and his Holeysphere adorns the introductions to each chapter of this book and also the design on the back cover.

Chapters 24, and 25, written by Slaivik Jablan and Ljiljana Radovic, both came about due to the work of the ethnomathematician, Paulus Gerdes, from Mozambique. Prof. Gerdes would visit tribal groups in Africa to study their folk art. As the result of these explorations, he found that the natives were using sophisticated ideas of mathematics to ply their trade. From this came the ideas of mirror curves and Lunda designs.

Finally, in **Chapter 26**, Jablan and Radovic study visual illusions and present us with the genesis of impossible objects and the beginnings of OpArt. This chapter introduces us to the work of several artists whose work were pioneering in the area of visual illusions.

Remark: After I completed this list of what influenced me in the various chapters of the book, it became apparent that the great mathematical puzzle master, recreational mathematician, magician, and revealer of mathematical curiosities, Martin Gardner, played a large role in this book. So, I would like to give a special acknowledgement to him for his contributions.

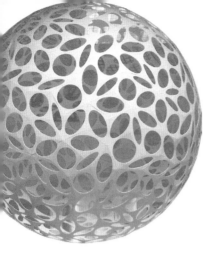

Index